Making Makers

AnnMarie Thomas

MAKER **MEDIA**
SEBASTOPOL, CA

MAKING MAKERS

by AnnMarie Thomas

Published by Maker Media, Inc., 1005 Gravenstein Highway North, Sebastopol, CA 95472.

Maker Media books may be purchased for educational, business, or sales promotional use. Online editions are also available for most titles (*http://www.safaribooksonline.com*). For more information, contact O'Reilly Media's corporate/institutional sales department: 800-998-9938 or *corporate@oreilly.com*.

Editor: Brian Jepson	**Indexer:** Wendy Catalano
Production Editor: Nicole Shelby	**Cover Designer:** Brian Jepson
Copyeditor: Kim Cofer	**Interior Designer:** Monica Kamsvaag
Proofreader: Jasmine Kwityn	**Illustrator:** Rebecca Demarest

September 2014:   First Edition

Revision History for the First Edition:

   2014-08-27:   First release

See *http://oreilly.com/catalog/errata.csp?isbn=9781457183744* for release details.

ISBN: 978-1-457-18374-4
[LSI]

For Sage and Grace

Contents

Foreword

The goal of Maker Faire is to "make more makers," so since 2006, we have gathered makers together and celebrated what they do. In 2013, there were 100 Maker Faires around the world, and a growing number of attendees are families. From the very first, we understood that meeting makers and seeing their projects inspired others to see themselves as makers. We invited everyone to get involved by creating and sharing projects based on their own interests and skills. While I knew Maker Faire would appeal to adults, I was surprised to see how kids were fascinated by the amazing variety of projects; they wanted to participate, and they wanted to become makers themselves. Sometimes, their own interest surprised their parents too!

At Maker Faires, parents were noticing how their children liked to play with cardboard as well as wood, how they liked to learn to use tools and how they liked to tinker. I suspect that some of these practices had fallen by the wayside—what seemed obvious and good to do for generations had been lost, forgotten, or just taken for granted. There is little hands-on learning found in classrooms. On playgrounds or outside of school, there is very little time for unstructured play. It wasn't clear how a child could become a maker without adult involvement. Fortunately, parents have began to discover that making is something to do together for fun while also being an authentic form of learning.

At Maker Faire Bay Area in 2014, I had a conversation about making and parenting that has stuck with me. Mike Neden, a professor of technology and engineering education at Pittsburg State in Kansas, was hanging out in the booth for Rokenbok, a 3D construction kit for kids. As parents and their children interacted with the kit and began building things, Mike found that parents asked questions about how kids learn, and more specifically, how they might learn to become engineers and scientists. Many of the parents asking questions had these careers and were working in Silicon Valley. Mike's comment to me was that these parents weren't sure how they themselves became engineers and scientists and were wondering how their own children might become them too. They understood the value of what they had learned to do—and how they thought—and they wanted to pass that on to their children. I've also met and talked to parents, particularly those like Julie Hudy, who we call Maker Moms. They recognize in their children the interest in and capability for making things. Because these parents are not makers themselves, they want to know how to support their children, whose experiences and interests are so unlike their own.

I am proud to publish this book by AnnMarie Thomas. It is a guide for parents who want to engage their kids as makers, but it is not a how-to guide or a self-help book. It offers insight into what we call a "maker mindset," which is a way of engaging with the world and gaining access to a toolset that is both mental and physical. The maker mindset can be used to change the world for the better. Dr. Thomas, a maker, a teacher, and a mother, helps us learn from makers so that all of us can foster the qualities we value most in them. It is clear that what many appreciate about making is not just the finished product, but also the character traits that the process helps develop.

Making is more than hands-on projects, but it starts with them. You can find lots of guides that tell you or your child how to learn electronics, woodworking, welding or sewing. You can find a variety of ways to engage children in building practical skills and developing original ideas. Yet what happens with a child's hands and what he or she creates are not nearly as important as what happens inside the child. This is what Piaget, a pioneer in our understanding of child development, called "cultivating the experimental mind" in his book *To Understand is to Invent*. He talked about a kind of active education that would *"lead the child to construct for himself the tools that will transform him from the inside—that is, in a real sense, and not just on the surface."* We can give our children access to tools and the time to practice using them to develop creative and technical projects. Yet, as parents who are makers, our most important project is helping our children become creative, lifelong learners. We need adults to facilitate these experiences and to create makerspaces in their communities that are accessible to all children.

Making more makers is what each of us can do, and what we must do together. These are the children who will make the future.

—Dale Dougherty
Founder and CEO, Maker Media

Preface

How are you going to design something if you've never built anything?[1]

— **WILLIAM GUILFORD (UNIVERSITY OF VIRGINIA ENGINEERING PROFESSOR)**

In the fall of 2011, as I was taking a break from grading my students' assignments, I found myself stopped by the preceding quote in an engineering education magazine. After six years of teaching undergraduate engineering design classes, I shared Dr. Guilford's sentiment. In my classes, I found that the truly innovative designs most often came from students who were able to couple rigorous analysis (which is the focus of many engineering programs) with a practical knowledge of how machines work. The latter is knowledge that comes primarily from taking things apart, putting them together, and learning what has worked (or not worked) in other devices. While I can assume that all of the first-year engineering majors that I teach have taken a math class, I can't assume that they've spent time taking things apart or building things. As someone who reveled in making things (out of wood, out of cardboard, out of fabric, out of sand, out of... anything) as a child and teenager, I had a hard time wrapping my head around the idea that many young adults, particularly those who were going into a field of study focused on creating things, had so little experience actually making things.

I read Guilford's article the same week that I attended World Maker Faire in Queens, NY. This event celebrated the creators of things ranging from robots to costumes. The engineering professor in me saw some incredible examples of technology, but more than anything, what I noticed were the people who were passionate about creating things and sharing their knowledge with others. The excitement was infectious and evident in attendees of all ages. Throughout the fairgrounds, there were opportunities for children and adults to learn skills like soldering, energizing conversations about the intricate details of various 3D printers, makers young and old showing off the things they'd made, and a pervasive air of curiosity.

---

1. Mary Lord, "Seeing and Doing," ASEE Prism, September 2011.

That sense of curiosity and collaboration is what I wish for all of my students, as well as for my own children.

I am definitely not alone in my desire to encourage children to actively create the world around them. Makerspaces, places where people get together to use tools and work on projects, are popping up around the world. We're seeing them in libraries, schools, community centers, and homes. Project instruction sharing sites (like Instructables and Make: Projects) allow users to freely share step-by-step directions for making everything from playful electronic gadgets to furniture to tomato soup. The Maker Movement, and the self-identified makers who are at the heart of it, are celebrating many of the qualities and actions that educators have long been trying to promote: lifelong learning, self-directed learning, communication, collaboration, creativity, and design. At a time when there is an increased emphasis on STEM (science, technology, engineering, and mathematics) in the PK–12 curriculum, the growth of the Maker Movement presents great opportunities for increasing technical literacy and reintroducing people of all ages to the arts of making and tinkering. The kids I watched at World Maker Faire, and at every Maker Faire I've attended, are asking great questions and doing "real" projects with the purest of motivations: they are curious and having fun. They aren't attending because it's a homework assignment, or learning to solder because it might be on a test. To be honest, most of them probably aren't even sure why learning to solder is useful yet; they just know that if they learn to do it in the Learn to Solder tent they'll get to wear, and keep, a cool blinking-light badge. But they'll leave the faire with more than just that badge. They'll leave knowing *what* soldering (or sewing, or woodworking, or cooking, or drop spinning...) is and that *they* can do it.

Which brings me back to Dr. Guilford's question: *"How are you going to design something if you've never built anything?"* I'd add to that and say *"How are you going to build something if you've never taken something apart?"* How are you going to come up with interesting ideas and solutions if you've never been allowed to play with physical and digital bits and pieces? It takes a playful, curious person to take things apart and imagine new ways to put the parts back together. This describes most makers, but also almost every young child that I've met. Anyone who has been around a roomful of children with access to a pile of craft or building materials has likely seen the happiness that typically accompanies such endeavors. Youthful creativity combined with readily available materials often leads to a whirlwind of wonderful things.

It usually doesn't take much effort, or the creation of any incentives, to convince young children to jump in and start making. As the age of the group gets older, though, the dynamic sometimes changes. We start to hear more questions: "Why should we do this?" "Am I doing this the right way?" I've made Squishy Circuits (a method for using conductive and nonconductive play dough to sculpt working circuits) with people of all ages, and I've rarely had a child turn down the opportunity to try it out. With adults, though, I've often seen reticence or protests of, "I'm not good at that sort of thing."

It's not coincidental that many authors who write about innovation, creativity, and design talk about the importance of approaching challenges with a childlike enthusiasm. Ursula Le Guin, an author who is known for her imaginative fantasy and science fiction writing, once worded this as *"The creative adult is the child who has survived."* Unsurprisingly, many people that I speak to about making share this approach. When I asked Amon Milner, a maker/educator who I will introduce you to in Chapter 5, what a "maker" was, he replied that *"[all] people are makers. And the conditions in which people can grow up and have that supported and still do it into adulthood is a very special person... Every [child] is a maker and some get to stay that way longer."* Perhaps, then, this isn't a book about *making* makers, but rather a book about how to encourage children to *remain* makers even as they grow up.

How do we empower children to become, and remain, makers? I think it's fair to say, particularly in the United States, that we're not doing such a good job at this. A 2009 study by Nuts, Bolts & Thingamajigs (the Foundation of the Fabricators and Manufacturers Association) polled U.S. teenagers and found that 83% of them spent less than two hours a week doing hands-on projects "such as woodworking or models," and 27% spent zero time per week working on such endeavors.[2] Interestingly, the same year, another study came out from the Kaiser Family Foundation finding that 8- to 18-year-olds spent an average of more than 50 hours per week on entertainment media.[3] While a lot can be learned through entertainment media, it sometimes seems like it is easier to find a cooking video game than it is to find a middle or high school that still teaches cooking. Are we giving children opportunities to create content, rather than just consume it? As Mitch Resnick, director of the MIT Media Lab's Lifelong Kindergarten group, points out, *"computers will not live up to their potential until we start to think of them less like televisions and more like paintbrushes... In my research group at the MIT Media Lab, our goal is to*

---

2. The Foundation of the Fabricators & Manufacturers Association, International, "Teens Turn Thumbs Down on Manufacturing Careers" November 16, 2009.

3. Victoria J. Rideout, Ulla G. Foehr, and Donald F. Roberts. "Generation M^2: Media in the Lives of 8- to 18-Year-Olds" (Henry J. Kaiser Family Foundation, 2010).

develop new technologies that follow in the tradition of paintbrushes, wooden blocks, and colored beads, expanding the range of what children can create, design, and learn."[4]

Once upon a time, spaceships resided primarily in movies, books, and the dreams of children, some of whom, after spending their teenage years working on their cars and tinkering, grew up to become the men and women who made manned—and unmanned—spaceflight possible. I believe that it is essential for us to empower today's children with the tools and skills they need to make *their* dreams tangible. The Maker Movement is a shining example of how we can do this.

[4]. Mitchel Resnick. "Computer as Paintbrush: Technology, Play, and the Creative Society," In Play = Learning: How Play Motivates and Enhances Children's Cognitive and Social-Emotional Growth, ed. D. Singer, R. Golikoff, and K. Hirsh-Pasek (Oxford: Oxford University Press, 2006), (2006): 192–208.

Safari® Books Online

Safari Books Online is an on-demand digital library that delivers expert content in both book and video form from the world's leading authors in technology and business.

Technology professionals, software developers, web designers, and business and creative professionals use Safari Books Online as their primary resource for research, problem solving, learning, and certification training.

Safari Books Online offers a range of plans and pricing for enterprise, government, education, and individuals.

Members have access to thousands of books, training videos, and prepublication manuscripts in one fully searchable database from publishers like O'Reilly Media, Prentice Hall Professional, Addison-Wesley Professional, Microsoft Press, Sams, Que, Peachpit Press, Focal Press, Cisco Press, John Wiley & Sons, Syngress, Morgan Kaufmann, IBM Redbooks, Packt, Adobe Press, FT Press, Apress, Manning, New Riders, McGraw-Hill, Jones & Bartlett, Course Technology, and hundreds more. For more information about Safari Books Online, please visit us online.

How to Contact Us

Please address comments and questions concerning this book to the publisher:

Make:
1005 Gravenstein Highway North
Sebastopol, CA 95472
800-998-9938 (in the United States or Canada)
707-829-0515 (international or local)
707-829-0104 (fax)

Make: unites, inspires, informs, and entertains a growing community of resourceful people who undertake amazing projects in their backyards, basements, and garages. Make: celebrates your right to tweak, hack, and bend any technology to your will. The Make: audience continues to be a growing culture and community that believes in bettering ourselves, our environment, our educational system—our entire world. This is much more than an audience, it's a worldwide movement that Make: is leading—we call it the Maker Movement.

For more information about Make:, visit us online:

Make: magazine: *http://makezine.com/magazine/*
Maker Faire: *http://makerfaire.com*
Makezine.com: *http://makezine.com*
Maker Shed: *http://makershed.com/*

We have a web page for this book, where we list errata, examples, and any additional information. You can access this page at: *http://bit.ly/making_makers.*

Makers

What is a "maker"? Quite simply, makers make things. Some makers build robots, some sew clothes, some prepare food, some design tools, some construct houses. "Maker" isn't a title conveyed after passing some test or degree program; rather, it is a self-identification. It's also not, by any stretch of the imagination, a new concept.

Humans have always been makers. Our survival is directly tied to our ability to create, or find, food and shelter, though we as a species shouldn't allow ourselves to feel too special because of this. From birds weaving elaborate nests, to beavers building dams, and spiders creating traps for their food, we are just like every other species in our biological need to make. What does set us apart, perhaps, is that we've reached a point where many people grow up without the ability to personally create any of the things (food, clothing, shelter) that they need for survival. Similar to its poll of teenagers mentioned earlier, Nuts, Bolts & Thingamajigs surveyed 1,000 U.S. adults in 2009, and found that 58% have never made or built a toy, and 60% admit to avoiding handling major household repairs.[1]

In the United States, the pride in creating things is such that President Barack Obama specifically mentioned makers in his 2009 inaugural address (*http://bit.ly/ obama-09*):

> *Our journey has never been one of short-cuts or settling for less. It has not been the path for the faint-hearted, for those that prefer lei-sure over work, or seek only the pleasures of riches and fame. Rather, it has been the risk-takers, the doers, the makers of things—some celebrated, but more often men and women obscure in their labor—who have carried us up the long rugged path towards prosperity and freedom.*

---

[1]. The Foundation of the Fabricators & Manufacturers Association, International. "Americans Don't Tinker Around with Hobbies, Home Repairs, Poll Shows," November 16, 2009.

I was delighted to hear this characterization of makers. Rather than focus on "eureka!" moments and successes, the president honors the hard work, lack of renown, and risk that is a more realistic portrayal of the road many makers follow. Yet for pride strong enough to make this something mentioned in a historic inaugural address, it seems worthwhile to reflect upon the apparent lack of knowledge among American citizens with regards to how things are made and why children should be taught to make things.

If humans have always been makers, why are makers and the Maker Movement garnering so much attention at this point in time? The story starts with a magazine. In 2005, O'Reilly Media (*http://bit.ly/orm-make*) launched a magazine titled *Make:*. *"The urge to make things is primal and unstoppable. In service of that universal urge, humans grab the tools and materials at hand—while a previous generation picked up a saw and bullnose rabbet plane, today's makers are likely to reach for a soldering iron and Cat 5 cable. Make: (http://makezine.com/), a new magazine from O'Reilly Media, celebrates and inspires those who are driven to make cool and unusual things with technology, for the pure fun of it."* From its start, *Make:* focused on showing the people behind projects, and the fun that could be had by jumping in and getting your hands dirty. The process of creating, not just the final outcomes, was as much of a focus as the technology.

The launch of *Make:* magazine happened while I was a doctoral student studying mechanical engineering and coming to the realization that my favorite things to do were teaching and designing/creating/tinkering. The week the magazine launched, I was at a conference also attended by Tim O'Reilly. He had left a few copies of the inaugural issue of *Make:* lying around and extended a generous offer for attendees to register for a subscription. The same day, I received a phone call from a department chair at the art school I taught at asking if I'd be interested in adding a robotics class to my teaching load. As I flew home from the conference, thinking of projects that would be fitting for a class of artists and designers, I was also flipping through the pages of *Make:* magazine.

When Issue 02 of *Make:* arrived, I was delighted to find instructions for Mousey the Junkbot.[2] I would go on to teach this project to many semesters' worth of Art Center College of Design students, and an intrepid group of 7- to 13-year-olds who took my robotics class at the Armory Center for the Arts. Here were projects that anybody with interest and access to some basic materials and tools could do. Watching my students building their Junkbots, personalizing them with everything

---

2. Gareth Branwyn, "Mousey the Junkbot," *Make:* Vol. 02, May 2005

from paint to custom shells, and then proudly showing them off to anybody who would look, I started to realize the power of what would very soon come to be referred to as "Making."

As I was finding in the classes and workshops that I was teaching, the final form of the artifacts, be they robots or clothing, wasn't where the power was coming from. Rather, it was through the community that was forming as people got together to make things, help each other, and then show off what they had made. For many makers, these endeavors were a hobby, not their main occupation. They'd spend their nights and weekends tinkering with technology, new and old, and taking to Internet forums and other outlets to get advice and show off what they were making. Typically, these forums were fairly specific and geared at certain skills or technologies. With the growth of *Make:* magazine and its related website, there was now a place where one article might be a profile of a "Live Steam enthusiast" who builds backyard steam locomotives, followed a few pages later by a description of an art installation involving *"500 stalks of chest-high, semi-flexible, fiber-optic strands arranged in a grid."*[3] By having an umbrella as wide as "people who make things," the magazine created opportunities to learn about a wide range of people and skills.

As readership of *Make:* grew, the concept of bringing the magazine to life, and providing an in-person way for makers to meet and share, arose. Founder Dale Dougherty (*http://bit.ly/nyt-dd*) recalls, *"Maker Faire started from the ideas in the magazine. We were covering lots of interesting people and I thought it would be interesting to bring them all together in one place. They did such different things, but they had a lot in common."* In its first iteration, Maker Faire brought more than 300 makers and 22,000 attendees to the San Mateo, CA, fairgrounds for two days highlighting the sort of ingenuity and projects that were enticing people to read *Make:*. Attendees could wander the grounds and meet makers working on a dazzling array of projects. While this initial attendance was impressive, what has happened since shows how excited the world is about a return to the celebration of "the makers of things." In 2013, the Bay Area Maker Faire had more than 120,000 people together, and 100 Maker Faires were held worldwide.

Even more exciting than the attendance numbers for Maker Faires is the composition of the attendees. More and more, families and children are making up a large part of the audience. So much so that the *New York Times* even asked "Is Maker Faire Made for Kids?" (*http://bit.ly/nyt-kids*) after writer Amy O'Leary attended World Maker Faire in Queens, NY, and found that children played an integral role

3. Arwen O'Reilly, "White Light/White Heat," *Make:* Vol. 01, January 2005

at the faire, both as attendees and as presenters. While the article went on to mention that some attendees thought this was a sign that it was a "less-edgy event," I see this as a sign that adults are realizing that this is something that they want their children to be part of. So much so that in the comments to O'Leary's post, attendees shared stories and videos about the positive experiences they had bringing their children to the faire.

As a parent, I brought my four-year-old daughter to a Maker Faire in 2012. (Her little sister got to attend her first Maker Faire at 10 weeks old, but I'm not sure it made much of an impression on her. My husband and I, on the other hand, learned that infants don't seem to like the sound of jet-engine-powered carousels.) While I was excited for her to take in the amazing projects, it was the makers themselves that I most wanted her to see. I wonder whether, years from now, she'll remember meeting the 11-year-old who creates her own maker how-to videos, or the man who painstakingly re-created landmarks out of toothpicks, and I hope that one day she is equally passionate about her own work and takes the time to share it with some other inquisitive little girl or boy.

Makers and This Book

The makers in this book were all born well before the launch of *Make:* magazine and long before "maker" was a word that regularly popped up in the press. None of them attended a Maker Faire as a toddler, or paged through *Make:* magazine as a high schooler. So how did they get started? What were they like as kids? That's what I wanted to know, and so I set out asking makers to tell me what they were like as kids. Over the past few years, I've visited and spoken with what at times felt like every maker willing to talk to me. I was amazed by the generosity makers showed as they gave up time from their busy schedules to reminisce about their families, their schools, their mentors, their early experiments, and what it was that made them excited as a kid. I heard stories of basement shops, but also of playing baseball. Some talked about their love of school, some had the opposite take. Many of the makers I interviewed are also parents. I learned from them about how they, as people who value hands-on work, parent their own children, and some of their doubts and concerns and challenges in that regard.

As I reflected on the discussions I had with these makers, there were some attributes that kept appearing in multiple people. While this isn't an exhaustive list, and while not every item on the list will apply to every maker, I do believe that the following attributes are ones that those of us who are interested in encouraging children in making should be thinking about (as you read the following chapters,

I encourage you to draw your own conclusions as well, and make your own list of lessons you can draw from these stories):

- *Makers are curious. They are explorers. They pursue projects that they personally find interesting.*
- *Makers are playful. They often work on projects that show a sense of whimsy.*
- *Makers are willing to take on risk. They aren't afraid to try things that haven't been done before.*
- *Makers take on responsibility. They enjoy taking on projects that can help others.*
- *Makers are persistent. They don't give up easily.*
- *Makers are resourceful. They look for materials and inspiration in unlikely places.*
- *Makers share—their knowledge, their tools, and their support.*
- *Makers are optimistic. They believe that they can make a difference in the world.*

To close this chapter, I'd like to introduce you to a maker whose story, both past and present, captures many of these attributes. In the fall of 2013, Jane Werner (Figure 1-1) was flying to Birmingham to give a talk to the board of the Institute of Museum and Library Services about making in museums. As executive director of the Children's Museum of Pittsburgh and a founding board member of the Maker Education Initiative, Jane has spent a lot of time thinking about children and making, which makes her a perfect person to talk to this audience. On the flight to Alabama, thinking over what she was going to say at the meeting, she found herself thinking about the clothing she had packed for the trip. It is fairly likely that Jane was the only person attending the meeting in a dress she had sewn herself. To her, this dress represented the role that making, and the empowerment that comes with it, played in her childhood. *"When I really think back on it, sewing gave me confidence and when you have confidence you feel you can learn, you can change things. I believe [sewing] was one of the reasons I became a museum director. It's the psychology of making. It's not just learning STEM, and not just learning the arts. It's learning about yourself."* Through sewing, *"this notion that I could change things in my world was driven home."*

As a child growing up in the small town of Hellertown, Pennsylvania, Jane and her older brother spent a lot of time outdoors making things: forts (*"We were the fort king and queen"*) and little boats to float in their creek. Indoors, Jane was also an avid maker. Her mother sewed and encouraged a young Jane to make clothing.

She became fascinated with fashion, and would peruse magazines at the library and then share the images with her mother who would say *"We can make that."* In this way, Jane began wearing clothes that were very much in fashion in big cities and magazines, but a bit less so in Hellertown. Her mother would help her find affordable fabrics at discount stores and in this way Jane became the first, and only, student in her high school to wear a midi skirt. "*They were big everywhere except for my town in Pennsylvania. I walked in and everyone looked at me like 'what are you doing?*" Even earlier, in middle school, Jane was hired by a friend's parents to make a dress for the girl's school concert.

Figure 1-1. A young Jane Werner, with her brother, wearing clothes made by their mother (photo courtesy of Jane Werner)

Jane has never stopped sewing. She started college as a fashion design major, but graduated with a synaesthetic art and education degree. After graduation, she moved to Pittsburgh, where she lived with her aunt who also sewed clothing. Her aunt, though, used really great fabrics and taught Jane that, in Jane's words, *"you really needed to invest some money and use Vogue patterns and look at things in a different way."* After her aunt moved away from Pittsburgh, at a time when Jane herself couldn't afford the nicer patterns and fabric, Aunt Judith would send patterns and fabric that she thought Jane would look great in.

Jane went on to work at science museums and launched her own exhibit design company. She later returned to Pittsburgh to become director of exhibits and programs at the Children's Museum of Pittsburgh. In 1999, she took over the executive directorship of the museum, which became known for its hands-on, exploratory exhibits with an emphasis on "real stuff."

In 2011, Jane attended her first Maker Faire at the urging of her friend, artist Ned Kahn. She remembers being amazed: *"I was blown away... This is everything that I've always loved. People doing whimsical, important, and playful things. Really interesting things."* She returned to Pittsburgh and decided to turn part of the museum into a makerspace.

The Children's Museum of Pittsburgh was, and is, created by the staff identifying things that were important in their own childhoods. MAKESHOP, the museum's makerspace, reflects this. Staff members reflected on things like sewing and messing around in the basement with hammer and nails. There wasn't a big budget for this project. Jane had $5,000 that she could allocate to the makerspace creation, which is an almost impossibly low sum for most museum exhibit design projects. The Education Technology Center at Carnegie Mellon offered to send three interns to Jane for the summer. Over the course of one summer, four people (the interns plus a museum staff member) turned the empty space into a room with saws, nails, and electronic circuits, and *"just started to mess around."* The space ended up so successful that it became a permanent exhibit. Jane and the other staff members realized that the MAKESHOP was a place where you could see interactions between generations, generating conversations and giving families an experience that they could build on at home. Families would return multiple times to the exhibit to work on multiday projects, and kids would spend hours at a time working on their creations. The day I was there, I watched a father and daughter working together on a colorful tank top using donated fabric scraps and a donated sewing machine.

While professionally Jane now spends more time creating situations in which her staff can be creative than in actually making things herself, she still sews. Laughingly, she tells me that when she's *"having a day when I'm at loose ends about [things]"* it becomes a *"fabric day."* She goes to her local fabric store and talks to Tammy, the owner, about fabric. *"I love the feel of fabric. I love that it drapes, and it's really fun to take something two dimensional and make it three dimensional."* While Jane no longer has to make her own clothes, she sees it as her creative outlet, as *"something that [she] can experiment with."* So much so, that she can often be found wearing items that she made herself.

Jane tells me about a recent phone call she received. It was a father calling with a question. He had taken his eight-year-old daughter to the Children's Museum of

Pittsburgh and they'd had such a great time sewing together that when he went home he went upstairs to their attic and took down the sewing machine that had belonged to his mother, who had died a few years earlier. Unfortunately, he realized that he and his daughter couldn't figure out how to thread the machine. He was calling to find out whether anyone could help. I suspect that he was a bit surprised when the executive director of the museum offered to teach them herself. Thus, this father and daughter returned to the museum where Jane taught them about thread tension and how to wind a bobbin. *"They were ecstatic,"* she recalls. The daughter immediately started thinking of the costume she wanted to create.

I love this story. An eight-year-old girl setting off to sew a Halloween costume of her own design, helped by her dad, using her grandmother's old sewing machine. She finds a mentor in a busy stranger, who was willing to share her time and knowledge (that she herself had put to use as a child to create clothes for herself and others). To me, that is the power of the Maker Movement.

Before leaving Jane to her busy day at the museum, I asked her what makers can learn from kids. I believe we adult makers can also learn a lot from watching children, as does Jane.

"The young kids don't care about failure. When they're 3 or 4, they just keep iterating and iterating. They just keep going. When we get them into schools, they start worrying about failing... you can watch kids... learn things quickly when they're making. You can see them making the connections. The joy in that... I think we sometimes forget about that joy in making things. It's so evident in kids."

My hope for all of us is that we begin to embrace this joy, and do our best to prevent our children from forgetting it!

2

Curiosity

Makers are curious.

They are explorers. They pursue projects that they personally find interesting.

When I was a child, I had an annoying habit of asking "why?" incessantly. It got so bad that once, at the end of a week-long camp, while the other kids were given titles like "best athlete" and "fastest swimmer" I was bestowed the name "Little Miss Why." In retrospect, I don't think it was meant as a compliment. As a parent, I'll admit that I've occasionally gotten exasperated when my daughters ask me that same question countless times in a row. I've even resorted to the unsatisfactory answer "Because!"

Based on what I've learned of other makers' childhoods, I definitely wasn't alone in my inquiry-based approach to life. Perhaps more than any other unifying trait in this book, curiosity seems to come up when makers discuss their childhoods. Makers are curious. Their youthful curiosity took on many different forms, but the common interest in gaining knowledge, skills, and stories seems universal. They wanted, and still want, to know *why, how, when,* and *what if?*

We send mixed signals to children about curiosity. After all, it's what killed the cat, so it can't be a good thing, right? I believe this couldn't be further from the truth. Curiosity, and self-driven pursuits, are behind all great innovations. Recently, I took a friend to her first Maker Faire. We walked around looking at projects and exhibits. She was impressed by what she saw, but wanted to know the motivation behind it. Why do makers make? Why would someone spend years building a city out of toothpicks? The projects that you find posted on Instructables, or in *Make:* magazine, aren't typically part of someone's job or a homework assignment. Rather, there is some internal motivation at work. Where does this come from? As someone who has seen amazing things come from these "just for fun" and "just because I'm curious" projects, my follow-up question is, "How do we nurture this in all children?"

Intrinsic Motivation

The makers in this book, now adults, still exhibit a childlike passion for learning. They find joy in doing new things and exploring new topics, skills, and places. Each of these are individuals who, even as grown-ups, see the world as full of wonder and possibility. They are driven by wanting to know, wanting to do. Few of them need to be incentivized to look outside of the proverbial box; it's something they're drawn to, regardless of what others around them are doing. Yet we, as a society, spend a lot of time focusing on external motivation for children.

Adults are very good at coming up with ideas for ways to motivate kids. Look at the proliferation of grades, badges, and awards. When it comes to teaching maker skills, I see this a lot: a competition to build the fastest car, or an assignment where your robot/windmill/circuit will be graded. I'm all for the existence of competitions and assignments that let students build and design, but I'm also interested in what inspires people to spend countless hours in the library, lab, garage, or computer room working on a project that no one has assigned to them. A project that, truth be told, may never work. A project with no due date, entry rules, or complete how-to instructions. Where does that passion come from and how do we encourage it?

As much as we might tell children to think outside of the box on a homework assignment, it takes a brave parent or mentor to let children go off script completely, to allow them to do something simply because they're curious. Doing so risks the possibility that the assignment won't be done on time or will be done in a way that might earn an unsatisfactory grade. Assignments are often created so that they can be completed in a certain amount of time and use a certain skill set. The world of open-ended, passion-driven projects is a bit harder to pin to a calendar or required parts list.

If we want to nurture intrinsic motivation, we need to demonstrate it. How often does a child meet a role model who is working on a project that he is truly passionate about? Who guides him on a project that he is doing because the process itself is enjoyable, maybe even more so than the finished product? My hope for my children is that they develop a spark that calls them to find their own interests, and to unselfconsciously throw themselves into projects that are meaningful to *them*. But, like any spark, these curiosities and passions are initially likely to be fragile and will need protection and kindling to grow.

Walt Disney once stated that *"[at Disney] we don't look backwards for very long. We keep moving forward, opening up new doors and doing new things, because we're curious... and curiosity keeps leading us down new paths."* More than anything, curiosity is the fuel that propels all makers forward. As an educator and as a parent, one of my greatest fears for today's children is that we are taking away opportunities for kids to get lost in their own curiosity.

Curious Kids

What does it mean for a maker child to be curious? When I ask this in the context of making makers, it seems to assume this means encouraging kids to take things apart. I'm a strong believer in the importance of letting kids take things apart, if they wish, but curiosity is about so much more than taking a screwdriver to your mom's iPhone. In fact, the first thing Dean Kamen, a prolific inventor, said to me during his interview was *"I wish I could tell you the myths about how as a three-year-old I was taking apart engines and electronics and televisions. Nothing could be further from the truth."* I also naively assumed that engineer Allison Leonard, whose "Machines Project" is an online repository of step-by-step photographed dissections of technology ranging from a Super 8 video camera to an Xbox controller, was a "take apart kid." It turns out that the Machines Project was motivated by the fact that she was *not* very interested in taking things apart as a child. *"The reason I did that project,"* she admits, *"is because I didn't do that as a kid. I think I always felt really late getting into engineering, into tech. I didn't get into it until my mid-twenties."* What she did do as a child was run around on the three acres of land her family had. That's where her interests, and curiosity, lay.

Dr. Lindsay Diamond (Figure 2-1), director of education at SparkFun Electronics, immediately answered yes when I asked her about this aspect of growing up. I began to imagine toasters and VCRs strewn about, but she quickly explained that, growing up in Florida they had *"an unbelievably large population of lizards. They were already deceased. I would take sticks and try to see what was inside."* Lindsay's take-apart subjects included *"all things biological,"* both plant based and animal. Flashing forward 20 some years, Lindsay is now a champion of open source education, particularly as it relates to electronics. Whether it's lizards or flashlights, Lindsay is among those educators promoting the importance of letting kids follow their curiosity.

Figure 2-1. Lindsay Diamond in her first laboratory, the family's backyard (photo courtesy of Marsha Levkoff)

School Days

Makers, young and old, want to learn. Thus, it seems like school would have been a pretty fantastic place for them. While some makers found school easy, very few makers spoke to me about loving school. Many told me about favorite teachers and classes, but others looked back less than fondly on their days in formal schooling. Through my work with the Maker Education Initiative, and through my research lab's focus on Playful Learning, I have been fortunate to meet incredible teachers. Teaching, particularly in the early childhood and elementary/secondary years, is one of the most important, and often, underappreciated, careers. Every week I meet teachers who invest every bit of time and energy they have into making learning come alive for their students. Unfortunately, more and more, I meet teachers who are becoming frustrated by the ways they are being forced to do their job. Not a single maker I spoke to mentioned joyful experiences taking tests or reading textbooks. They did talk about teachers who let them do projects, or librarians who found intriguing books for them, or stories and field trips that their educators brought to life for them. What breaks my heart is knowing that so many of those experiences are becoming rarer and rarer in today's education system.

An eye-opening discussion on this topic was one that I had with Steve Jevning (Figure 2-2), founder of Leonardo's Basement, a youth makerspace with locations in both Minneapolis and St. Paul. I long knew Steve as an informal educator, having served on the Leonardo's Basement advisory board. What I didn't know was that Steve's original plan was to be an elementary school teacher. Steve was born in 1954 in a small town in Minnesota. Rather than playing with toys, Steve spent his days finding materials, natural and manmade, to create with. As he put it, *"you had to build stuff to make the environment more fun to play in."* His maternal grandfather, a minister, had a woodshop in his basement and from the age of four or five, Steve was welcome down there. (He still remembers, though, that the lathe was off limits.)

Figure 2-2. Steve Jevning as a child (photo courtesy of Steve Jevning)

Steve's other grandfather was a farmer, and Steve remembers his farm as being like *"a giant playpen with tools and machinery and haylofts and ropes to swing from."* He would explore the neighborhood, borrowing equipment from construction sites, and visiting the local blacksmith shop where he'd watch in awe as the workers would fix things using mechanical bellows that they had made themselves. In short, Steve was a kid who was thirsty for knowledge about how to make things. Through his fascination with building models, he taught himself things like scale and proportion. No one assigned this to him, or graded him on it.

One thing that bothered him then, and now, was the way that he and his friends were tracked based on grades, rather than interests. *"In that period of time you didn't really know what your friends' interests were because it was so predetermined by the adults. That's why Leonardo's Basement became a kid focused place because I so objected to those learning experiences where adults knew best and you were never asked what you wanted to do. I learned at an early age the value of motivation."* Steve found high school industrial arts classes intriguing, but was regularly questioned by others about *"are you going to college or are you going to be just a shop guy?"* Many of his friends, who had high grades in school, had to fight to be in shop class. They were pushed to stay away from those classes, and to take more "academic" classes instead. Steve considers it fortunate that he was a C student, because that meant he could go either way, taking more academic classes or taking industrial arts classes.

Steve had relatives who were teachers so he chose that field when he eventually went to college, after a few years of traveling around the country as a carpenter. As an education student in the '60s and '70s, there was lots of discussion about experiential learning, and Steve was hopeful that he could really make a difference in how things are taught. He spent most of his undergraduate time in student teaching and independent study classes, and gained a degree in elementary education. He reveled in writing curricula that he thought would be more engaging *"and not as lame"* as the textbooks they were using at the time. Exposed to science kits, he found himself thinking *"This is so dumb. There's nothing in this kit that you can't just get at the local grocery store or hardware store. Let's go to the grocery store, let's go to the hardware store, buy these things, and then use real stuff and do real experiments, rather than sit at little tables with plastic spoons."* Steve is quick to note that this was the era when you could still walk into a store in any city and buy a real chemistry set with real chemicals. That was his first *"inkling that real things had more educational value than pretend things. That making something yourself and finding an existing product made more sense than using someone's kit that was a representation of that."* Steve had become an education major thinking that he could change education, but became frustrated when he felt that the system fought against the sorts of methods he wanted to use. He valued exploration, and playing. *"I just knew that each kid was different and the goal of education should be to figure out how to help each individual find their own way.*" To him, individualized learning, and allowing kids to find their own interests, was critical. After two years of student teaching in a variety of school types, he decided he wasn't going to find a "place" for himself and left the world of formal education. He became a carpenter again and built houses.

When his son was born, Steve became a stay-at-home dad, who later would do construction work while his son was in school. At the end of the school day, his son and friends would go to the Jevning household to build things. Steve would eventually start a science club at the local elementary school. His belief was that *"instead of buying toys, you can open the cabinets [in] your kitchen and let [the kids] explore."* Steve created a variety of opportunities for kids in his neighborhood from teaching school clubs and leading inventor's fairs. The growing interest in what he was doing led him to found Leonardo's Basement, a workshop where children, adults, and families can tinker and create with real tools and real materials. True to his educational philosophy, Leonardo's Basement is about relationships. Steve puts effort into making sure that Leonardo's Basement is a place where young and old learn and work together. *"Just because you are older doesn't mean that you can teach someone who is younger, or that you want to, or that you should."* It may not be the way he originally planned it, but Steve was able to fulfill his vision of individualized, student-driven education, involving *"real stuff"* and *"real tools."*

The question more and more educators seem to be asking today is why they can't do more of that in their classrooms. It's a great question, and I believe that all of us should be supporting those teachers who are trying to bring this sort of learning back into the school day. We all need to ask whether we are giving today's children the skills and knowledge that they will need to accomplish those things that need to be accomplished in the decades ahead. Steve strongly feels that one of the keys to this is that *"Finding something that you give a hoot about is just a really big part of making it through the tough times, but also enjoying the good times."* I agree with him, and hope that we're allowing all children to find just what it is that they give a hoot about! Steve proudly notes that 15 years after founding Leonardo's Basement, he is regularly being asked by schools to help them design resource rooms for designing and building projects.

Powerful Stories

For many of today's adult makers, books and reading played a critical role in helping them find the information they desired. Nearly every maker I spoke to could immediately list books and magazines that they read, sometimes more than three decades later. Some gathered up how-to books, while others read science fiction novels. Some read classics, while others became enthralled by nonfiction. I did note, with some amusement, that not a single maker listed a book that had been assigned for school. Rather, the childhood books that they still recall fondly were the ones that they found, and chose, for themselves.

One of my favorite discussions about reading was with Dean Kamen. While many interviewees talked about school having been easy for them, that wasn't the case for Dean. He cared about learning, but didn't turn in the homework or do well on the tests. He found himself frustrated by textbooks and school because they spent too much time teaching things that were already known. He wanted to learn the things that weren't known. Reading was, and still is, something that Dean was slow at.

As he put it:

> I'm a very, very slow reader, so reading long novels... I don't have the time. It's too much work for me. But once I realized that they called it Newton's law because of this guy Newton, he must have been a genius, he must have written a book. Oh! He did! Principia! Galileo wrote books too, Two New Sciences. Archimedes' principle: what did Archimedes write? So I figured, you come up with these great names, go and find out what they wrote. Instead of trying to learn what they did. A great genius spent his whole life writing Principia, and we're supposed to read one paragraph about it? I'm no genius, I'm no Isaac Newton, but somehow I'm supposed to learn and understand and appreciate his work, from one paragraph in an elementary school book that talks about Newton on one page, and the next page is how to put pins in frogs, and the next about what electricity is? No. I'll go get what that guy Newton wrote and if it's a whole book it'll take me a week or a month to read it, but when I'm done at least I think I'll understand it or I hope I'll understand it. I just decided I loved reading, very slowly, by myself, what these great people did and from it I developed an understanding of, and an appreciation for, the science, for the math, for the engineering that I think is just a beautiful piece of life that most people never get to see.

Sometimes those books loved in childhood offer hints to later projects. Chris Anderson (Figure 2-3), former Editor in Chief of *Wired* and founder of 3DRobotics (maker of aerial robots), regularly took books about aeronautics out of his school library. Growing up before the age of online shopping, Chris would also pore over mail order catalogs. Given his youthful lack of substantial spending money, most of his catalog reading was aspirational. Occasionally he came across something in his price range. Often these potential purchases could be found in the back pages of magazines such as *Popular Mechanics*. Twelve-year-old Chris's big purchase was for what he thought was a submarine kit. Full of anticipation, he awaited his submarine kit, dreaming of adventures that could be undertaken when the vehicle was finished. In actuality, what he had ordered was a set of plans. Thus there was understandable disappointment when Chris opened the package to find a set of *"badly copied blueprints."* Not only that, but the first step in the plans was to acquire a surplus U.S. Air Force P-51 external fuel tank. Not an easy task for a suburban high schooler. *"It was really kind of discouraging. Those were the days when dreams were shattered every time the packages arrived in the mail."* Sadly, he was unable to find a spare fuel tank in his suburban neighborhood, and his hopes for having a personal submarine were dashed.

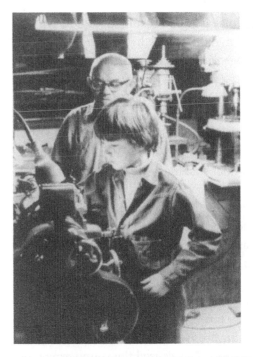

Figure 2-3. Chris Anderson machining with his grandfather, Fred Hauser (photo courtesy of Chris Anderson)

Chris went on to lament that he thinks he was born too early. He didn't get his first computer until he was 18, and still dreams of what he would have done had he gotten one when he was 12. This raises an interesting point. Quite a few of the makers that I interviewed commented on how computers didn't play a large role in their early childhoods. Many wistfully discussed how great it would have been to have had one. I wonder if they would have grown up to be the makers that they are today had they had easy access to computers as a child. I pointed out to Anderson that had he had a computer, he might not have had his experiences with sitting in his room reading submarine construction plans. Sure enough, he remembered that *"The moment I saw a computer I pretty much gave up on anything mechanical on hardware entirely, and I just programmed for the next 20 years. And that's what happened, it's incredibly seductive and so maybe that kind of pre-computer period gave me an appreciation for the kind of mechanical side of the equation that's driven me to the makers world today."*

Following Curiosity Wherever It May Lead

It was interesting to note that some of the makers that I assumed would have had highly computer-influenced childhoods often didn't. Nonetheless, *curiosity* continued to be an oft-mentioned thread in these stories. India, Noah, and Asa Hillis are the children of Danny Hillis (see Figures 2-4 and 2-5). Danny is a pioneer in the field of computer science, particularly parallel processing, and we'll look at his own youth in Chapter 8. Given that Danny founded a computer technology company prior to the births of his children, I expected his children (all talented makers) to tell stories of a high-tech childhood. Instead, I was regaled with tales of rabbits, treehouses, hand tools, and low-tech mischief making. Also, lots and lots of exploring, of both places and of topics.

When I asked what drove them as children, all three Hillises explicitly mentioned "curiosity." Noah and Asa, twins now in their twenties, have fallen into the "take things apart" category for as long as they could remember. They recalled a time when they, as toddlers, managed to take apart their crib and, subsequently, their window's locks. Much to the horror of their mother and neighbors, this particular incident led to them ending up on the roof. Thankfully they were also fascinated by cameras and having their picture taken. When a frantic neighbor rang their house's doorbell and announced that the boys were on the roof, she was given a camera and was told to take the boys' picture. This caused the young explorers, who were also intrigued by photography, to freeze and pose while their mother ran upstairs and plucked them back into their bedroom.

Figure 2-4. India Hillis an an early woodworking project (photo courtesy of Danny Hillis)

Figure 2-5. Asa and Noah Hillis exploring (photo courtesy of Asa Hillis)

Introduced to a variety of tools as children, they have a long history of purposely seeking out ways to learn new things, and their parents were quick to encourage their interest. As Asa put it, *"If we were interested in something, [our parents] made it happen."* Their mother, Pati Hillis, was trained as an artist, and passed her love of art onto her children. She also made sure that the children had roles in the design and construction of their family home—ranging from counting doorways to helping lay the tile in the courtyard. When they expressed a desire to learn something, Pati would find them tutors.

Figure 2-6. The Hillis family collaborating on a castle (photo courtesy of Danny Hillis)

Rather than simply dabble in a variety of topics, the Hillis kids immersed themselves in their interests. Prior to his college experiences in furniture design, Asa had already learned wood and metalworking, guitar building, and basic furniture upholstery, but knew little about electronics. His solution? Signing up for an entire year of electronics classes, despite it not being a required part of his major. The twins' love of hands-on craft and artistry even foiled an early opportunity for them to pursue more digital topics. When he was 12, Asa briefly had an early iPod Video. He quickly sold it to one of his teachers, and used the money to buy a small wood lathe that he learned how to use, eventually becoming skilled at creating wooden pens. To help find buyers for the pens, Pati Hillis would bring her son's creations to conferences and gatherings. This turned into a small company where he employed his twin and equally youthful neighbors, running a wooden pen making enterprise on the weekends.

The link between Noah and Asa's childhoods and their current furniture building seems rather logical. When I asked about topics they loved to study as a child, India answered for the trio by noting that they *all loved math. We had really incredible math teachers.* Similarly, when I asked about toys, she noted that they didn't get toys, they got *"cedar blocks."* Woodworking, and related endeavors such as treehouse building, showed up in all three of their narratives. Noah summed it up like this: *"When we were kids we had freedom to run into the shop to build whatever."* These days, wood and mathematics are an omnipresent part of the boys' lives: *"[as a child, Math] was like a puzzle. There was a clear answer most of the time. It goes back to what I'm doing now, there's a lot of math in furniture design. Whether you know it or not, calculating an angle or how much stress will be on a certain joint, it's all problem solving."*

We don't get to pick what our children are curious about. We do, though, have the opportunity to guide them as they nurture this curiosity. I asked Chris Anderson, the young submarine builder turned aerial roboticist, whether he involves his five children in his making projects. His reply? *"Sure, I spend every weekend coming up with projects that they'll want to do, whether they're school projects or just fun things... Every weekend I push this, and every weekend it's like pulling teeth."* He admits that his desire is partly a *"selfish"* attempt to mold his children in his own image. And while he may not, as of yet, have been successful in his attempts to lure his children into the maker world as deeply as he is, I can't help but notice that he's doing something just as, if not more, important. The Anderson children are growing up watching their father tirelessly pursuing his curiosities, and his belief that science and technology are one of the pathways to making a positive impact in the world. Only time will tell which pathways they will choose to make their own marks.

More Tape!

Many parents, myself included, seem to have aspirations for what their children will find fascinating. Lisa Regalla, of the Maker Education Initiative, recounts a story of a father who came up to her after hearing her speak about the importance of making. He mentioned how despite his and his wife's efforts to get their sons interested in other materials and undertakings, the boys insisted on cutting paper and taping it together. It had reached the point where the parents were considering getting rid of the scissors and tape altogether. However, after listening to Lisa speaking about young makers and ways to support them, he had a new idea. Rather than get rid of the tape and scissors, he was now heading home to get *"more paper, and different types of tape!"*

We don't get to pick children's interests, be it paper and tape or submarines. That's the beauty of passion: it's personal. Imagine the horror of a world where everyone wants to know the same things. Even more horrible, imagine a world where we are told what to care about and are only allowed to pursue those endeavors. One of the most exciting things about life is how different we all are. I can assign the same design challenge to 30 students and end up with 30 (or more) unique solutions that are colored by the life experiences those students have all had. Where one of them sees airplanes, another may see birds, and yet another is dreaming of acrobats. The greatest gift we can give today's young makers is our support as they let their own curiosity take flight.

3

Playfulness

Makers are playful.

They often work on projects that show a sense of whimsy.

Play is a natural consequence of curiosity. Sophi Kravitz grew up playing unusual games. Games like "Where is Patty Hearst?" and one in which the periodic table of the elements was the board. In that game, you would roll the dice and then recite from memory all the elements, that came before the one that you landed on. You would then be quizzed on aspects of that element such as its atomic weight or common applications. The periodic table game was a favorite of hers, and she would eagerly play it with her two younger brothers. Needless to say, these weren't games that were purchased from a local toy store. The games were the handiwork of their father.

Sophi and her three siblings grew up with fairly strict rules about how to spend their playtime. There was no television in the house, and her parents were careful about which toys were allowed. To avoid falling into gender stereotypes, Sophi was given trucks and her brothers were given dolls. Amusingly, this strategy failed as her brothers would throw the dolls out of the baby carriage and replace them with Sophi's trucks. Sophi, on the other hand, began to put clothes on the trucks and put them to bed. Without access to dolls, Sophi simply designed clothes for everything around her, be it trucks, pine cones, or sticks. Games, stories, and imaginary worlds were an omnipresent part of her childhood: *"I could go in my head, and I can still go into my head, and make up anything."* She read constantly, and reveled in fiction. At her home she was allowed to read anything that she wanted, which occasionally led to awkward situations at school. In fourth grade she chose a rather explicit book to read out loud to her classmates, which led to the teacher calling home. Her mother informed Sophi's teacher that their family didn't believe in banning books, but also advised Sophi to keep such books at home. For years, she added to a *"never-ending book"* that she wrote about a group of friends and their adventures.

Sophi's parents stressed the importance of formal education, but school was not something that she was passionate about. When she needed to she could easily get a perfect score on finals, but she was more interested in stories and creating things. She eventually tired of high school, and managed to graduate early at 16.

She took full advantage of the loudness of 1980's fashion by piling hats with objects and using Christmas ornaments to embellish clothing. Over the following years, she studied topics such as fashion design and sculpture at FIT and SUNY Purchase. Nothing quite fit what she wanted, and her parents were adamantly against her becoming an artist. They wanted her to pursue science.

While looking for employment, she came across an ad posted by a sculptor looking for an assistant. When she showed up at Aesthetic Creations to interview for the position as Neal Martz's assistant, he showed her one of the goriest props that he had made for *Silence of the Lambs*. Sophi didn't flinch while looking at *"this nasty, bloody thing,"* which seemed to impress Neal. He then asked to look at her hands, pronounced them "dirty" (as she had been silversmithing all day), and hired her on the spot.

Little did Sophi anticipate that her first project would entail her poking hairs into the stomach and buttocks of a life-sized prop (Figure 3-1). She and the other assistants were given the task of making these hairs look as realistic as possible on a life-sized dummy of Howard Stern that was used in the movie *Private Parts*.

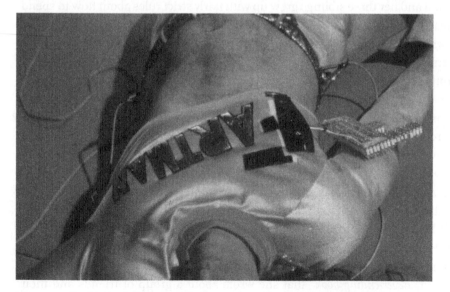

Figure 3-1. Sophi Kravitz's first movie prop project (photo courtesy of Sophi Kravitz)

Eventually, Sophi's parents came to visit her in her studio (Figure 3-2), where they found her surrounded by realistic looking body parts, and gallons and gallons of fake blood. The little girl who grew up without a television, lost in fantasy and stories, and memorizing the periodic table, became a woman who could use her knowledge of science to bring stories to life.

Figure 3-2. Sophi Kravitz in her electronics lab (photo courtesy of Sophi Kravitz)

These days, Sophi has moved on from special effects. She now has her own engineering consultancy, where she works on a wide range of projects, most commonly focused on industrial machine design. When I ask her to speculate on why she's been so successful in this line of work, she muses that these companies are looking for help with innovation and creativity. *"That's what we do as makers. We think of things that haven't been done. For me originality is key. I don't want to do projects that have already been done."* She has been coming up with original and creative ways to entertain herself and others her entire life.

An Invitation to Play

David Lloyd George, the early 20th-century British politician insisted, *"The right to play is a child's first claim on the community. Play is nature's training for life. No community can infringe that right without doing deep and enduring harm to the minds and bodies of its citizens."* Of the many statements about the importance of play, this has stuck with me the longest. Play allows our children (and us as well) to choose the activities we find enjoyable without worrying about outcomes. Play is a critical element of the human experience. We play games. We play music. We play sports. We play house, and fairies, and make-believe. Children learn about the world through play. That said, we have a complicated relationship with the concept of play. We talk about the importance of playing, but we also use phrases like "Stop playing around and get to work," implying that time spent playing is of a lesser value. How often do we, as adults, truly allow ourselves to focus on something we're doing "merely" for fun, and not worry about how it will turn out? Similarly, how often do we really give our children the time, and space, to freely choose how they spend their time?

One of the most striking things about visiting a makerspace or Maker Faire is how much fun everyone is having. When Dale Dougherty talks about why he started *Make:* magazine, he often talks about makers as being people who see new technologies as an invitation to play. Of course, just because something is playful doesn't mean that it is also easy. As an educator, I have always found that one of the tricks to getting kids (of any age) to work really hard is to help them have fun while doing it. Scholarly works on play, such as that by Fergus Hughes, define play as having elements including *"freedom of choice," "personal enjoyment,"* and a *"focus on the activity as an end in itself rather than on its outcomes."*[1] These criteria map nicely to what we're seeing in the maker movement. Most makers make because of their own personal interests and the pleasure of bringing their ideas to life. Similarly, many makers' projects are never truly "finished," but rather in a constant improvement process.

Sophi's story is an example of someone who has been playful her entire life. This sense of play, which she infuses into almost every aspect of her life, allows her to be creative and to see things differently. Deservedly, she prides herself on her ability to approach problems from new directions and see things that others don't.

Playing with Numbers

Almost any discussion of the Maker Movement in the United States, particularly as it relates to education and children, eventually meanders through the MIT Media

---

[1] Fergus P. Hughes, "Spontaneous Play in the 21st Century," *Contemporary Perspectives on Play in Early Childhood Education* (2003): 21–39.

Lab's Lifelong Kindergarten group (*http://bit.ly/llk-group*). LLK, as the group is known, was the birthplace of the wildly popular Scratch programming environment for children, LEGO robotics kits (including LEGO Mindstorms), Drawdio, MaKey MaKey, GlowDoodle, and countless other devices and methods for introducing children to the joys of being creators. The founder and leader of this prolific lab is Mitch Resnick (Figure 3-3).

Mitch worked on a number of memorable projects as a kid. For example, at one point, when he was in elementary school, his parents let him turn the backyard into a miniature golf course. That said, he doesn't consider himself to have been very interested in the physical making of things as a kid. Rather, you'd be more likely to find him doing a book of math puzzles. Mitch loved puzzles, particularly analytical puzzles like logic puzzles. This was how he played, entranced by algorithms and numerical manipulation. He saw this as his identity.

Now in his 50s, Mitch still vividly recalls the joy that learning math brought him. One story he recalled was from second grade. His sister, two years older, had just learned to multiply multiple-digit numbers, such as 12 times 33. At the time, Mitch knew about multiplication, but only for single-digit numbers. After his sister showed him this next step, he immediately asked her *"Does anyone else know about this?"* sure that this must be a secret only a few people knew. Otherwise, he assumed, everyone would be talking about it! Looking back he recalls that as the moment when he first recognized the *"beauty of an algorithm."*

Figure 3-3. A young Mitch Resnick (photo courtesy of Mitch Resnick)

Young Mitch also enthusiastically participated in sports. He played baseball, basketball, and tennis throughout elementary and middle school. He quickly realized that his athletic pursuits could also have a computational component to them, and he became interested in baseball statistics. His father showed him how to use the newspaper's listing of scores, batting averages, and standings to analyze games. This new way of playing with numbers enthralled Mitch, merging his passions for sports and math.

For a young Mitch, mathematics was his tool of choice for exploring the world. As he got older, Mitch's toolset broadened. In 10th grade, he was introduced to computer programming because of his strength in mathematics. In short order, he created his first program, a poker game. As an undergraduate at Princeton, though, he chose not to take any computer science classes. At that time, he only saw computers as a means of getting things done. That all changed when he met Seymour Papert, one of the pioneers of computers and learning. *"Seymour was a maker at heart, and he saw that the computer was a new tool that let kids be makers."* Mitch's eyes were opened to new ways that computers could be used. Mitch's very first graduate school project? Connecting LEGO to the computer. Mitch and a fellow student worked on ways to allow kids to control LEGO creations from a computer. This "programmable brick" work was instrumental in the development of LEGO Mindstorms. Thus began a collaboration between Mitch and LEGO that continues to this day, and the beginning of Mitch's career creating ways for children to play with physical and digital tools simultaneously.

Mitch never gave up on his athletic pursuits, and still plays a weekly tennis game with a friend, and the sheer joy of the seven-year-old learning about multiplication is still present. After each week's tennis matches, Mitch records all of the games' statistics. At the end of the year, he and his tennis opponent, a biologist, see who came out ahead in wins. One year, Mitch noticed an asymmetry in the data: his opponent had won 70% of the sets during the year, but only 56% of the games. Trying to understand this difference, Mitch wrote computer simulations (using his Scratch software) to analyze the information. Meanwhile, the biologist started to explore the question using the tools of his trade, using equations to model their matches. Reflecting on this, Mitch sees it as part of the maker mentality. By creating simulations and equations to model their activity, both tennis players were *"making something to explore and understand better."*

Earlier, I quoted Jane Werner's observation that while adults often seem to have forgotten the joy in making things, children have not. One of the reasons that I cherish time spent with Mitch and his students is that these are clearly adults who haven't forgotten the joy in making and in playing. Most people's vision of a lab at MIT involves lab coats and sterile surfaces. Mitch's lab, on the other hand, is bright, cheerful, often noisy, and full of brightly colored toys ranging from LEGO parts to craft clay. At a school often known for its intensity and its students' high stress levels, Mitch has created an environment where play, and children, are respected. *"The Lifelong Kindergarten group is sowing the seeds for a more creative society. We develop new technologies that, in the spirit of the blocks and fingerpaint of kindergarten, engage people in creative learning experiences. Our ultimate goal is a world full of playfully creative people, who are constantly inventing new possibilities for themselves and their communities."* The result? Tools such as Scratch, which has more than one million users.

Playing with Sound

These days, graduate student slots in Resnick's Lifelong Kindergarten group are highly sought after. So what does he look for in successful applicants? He looks for students who worked on interesting projects. Notably, he doesn't look at their grades or test scores. *"I look at whether they are going to be good community members."*

Not surprisingly, the individuals that he chooses to let join this community are incredibly interesting, talented, makers.

One of the innovations to come out of Mitch's research group is the "Creative Learning Spiral," which describes the learning methodologies that LLK's projects use. It was joked that each of the five items in the spiral (Imagine, Create, Play, Share, and Reflect) represented one of the students in the group. So who was rumored to be Play? That would be Eric Rosenbaum (Figure 3-4), who stated that play is *"what I'm drawn to. It's how I spend my time. How I express myself."* Eric is someone who has spent his entire life finding unique ways to play, and lately, inventing new forms of play for everyone.

When I think of "play" I often think of laughter and carefree endeavors. Thus, I was a bit surprised to find that Eric had been, by his own description, an anxious child. He wanted to know what was going to happen before it happened. While a bit fearful of the unknown in the real world, as a child Eric loved, and still loves, *"imaginary worlds full of unexpected magic."* Amusingly, when I asked this self-described former anxious child about his strengths, he was quick to discuss flexibility and improvisation. The irony of this isn't lost on him. *"I value in myself the ability to think flexibly, to turn things inside out or upside down to find a new direction."*

Figure 3-4. Ten-year-old Eric Rosenbaum (photo courtesy of Eric Rosenbaum)

Eric is drawn to generativity, to new and surprising ways to combine things. This takes many forms: from inventing games, to making improvised music, to building creative software and tangible construction kits. As an elementary school student, he often spent long creative afternoons with a friend named Elan, who lived just up the street. They were constantly making up new games to play. Some involved chasing each other with stuffed animals, others involved running up and down the stairs or dueling with Wiffleball bats, blankets, and laundry hampers. Often they invented elaborate rules for games to play with dice. One was a baseball simulation that required several weeks just to play through nine innings. Another was an adventure on a dangerous planet, complete with character sheets (including "alien biologist" and "ninja"), pages of rules, a map on graph paper (Figure 3-5), and numerous colored pencil illustrations of aliens with laser guns (Figure 3-6). They also modified existing games, such as using wooden pattern blocks as pieces on a chessboard.

By the end of middle school, Eric had progressed with a group of friends from role-playing games, with rulebooks and dice, to a game focused on storytelling: the collaborative creation of a fantasy world. They made maps and illustrations, lit candles, and built props from cardboard, wood, and wax. Of course this was all done while hidden away in a basement, away from the eyes of their peers, because it wasn't cool. These secret basement story games played a crucial role in developing Eric's imagination.

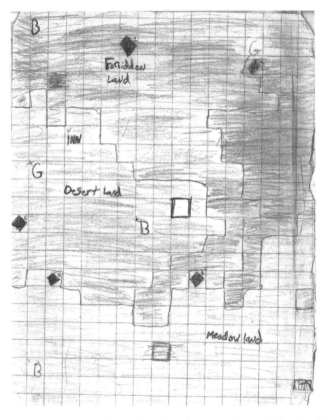

Figure 3-5. A map from the "Aliens!" game that Eric developed with his friend Elan (drawing courtesy of Eric Rosenbaum)

Figure 3-6. Creatures from the "Aliens!" game that Eric developed with his friend Elan (drawing courtesy of Eric Rosenbaum)

Unlike Mitch Resnick, Eric was not an early fan of mathematics, at least in school. *"I could perform arbitrary manipulation of symbols but I wasn't happy about it."* It seemed meaningless to him. High school classes in jazz improvisation, on the other hand, made a profound impact. Eric had fallen in love with the trombone the moment he heard it play "When the Saints Go Marching In" at a demonstration at his elementary school. This, along with the jazz records his father would play at home, led to a lifelong passion for music. He hid the role-playing games from most of his friends, but music was a creative world that brought with it a whole new social environment. He could express himself, and share his sense of humor, by, for instance, choosing the name for a piece performed by his middle school jazz ensemble ("The Icy Cold Hand of Death Blues"). More importantly, he could develop a sense of confidence in performing and improvising, which requires adapting to the unexpected.

I can't help but notice that the playful activities that Eric partook in seemed to be almost completely without competition. The role-playing games and musical endeavors were an end in themselves—they were never about winning or losing. In further discussions with Eric, it is evident that this is a running theme in his approach to play, even as an adult. He is wary of using assignments or competitions to motivate makers:

> *Assignments involve grades, which teach you that an external authority is the source of value judgments. We need to teach kids that they get to decide what is valuable... authority is there to inspire or reject, as needed. Competitions are even worse—they cause the same external judgment problems as grades, but they also teach 99% of people that they are losers because they didn't win. They also imply that what you do is only worth doing if you're the best, which is very harmful. I think that competition really impaired my musical life in high school... I'm happier now with my actual skill level and the value in what I can do with it.*

Play as Work, Work as Play

These days, Eric is finishing up his Ph.D. dissertation about learning music through play. His goal is to create new pathways into musical creativity that allow people, young and old, to construct their own musical instruments and compositions. To say that his work is having an impact seems like an understatement. His MelodyMorph app lets users develop, and tinker with, their own interactive instrument layouts, choosing notes, modifying timbre, and customizing to their own preferences. Another app he created, Singing Fingers, allows children to paint with sounds, creating digital, interactive, musical paintings. More recently he has begun collaborating with visual artists on physical paintings that allow you to trigger melodies by moving a hand across the painted surface.

The most well-known of Eric's projects, though, is the MaKey MaKey, developed with fellow graduate student Jay Silver. The MaKey MaKey is an open source circuit board that allows you to turn everyday objects, such as plants, fruit, or friends, into computer keys. This construction kit has allowed teachers to encourage their students to tinker, and see their world as a construction kit. But it is being used in all sorts of other contexts as well. Assistive technologists can use it to build custom interfaces for people with limited mobility, such as a game controller made from cardboard and Play-Doh, strapped to a wheelchair arm, that enabled a boy to steer a car in a racing game for the first time. Musicians are using it to extend the capabilities of their existing instruments, and invent entirely new ones for performance in music videos or on stage. Graduate students, artists, designers, ad agencies, museum educators, software hackers, and a wide range of others are using it as a new way to imagine and create. *"The tools that I'm making are about bringing that experience, that generativity, to other people. Making it possible for other people to experiment and generate new ideas. And that's what drew me to Lifelong Kindergarten in the first place."*

Many of the adults that I interviewed for this book have crafted paths for themselves where it is almost impossible to tell when they are working and when they are playing. When I asked Eric what he does for fun now, as an adult, he started talking about a recent musical jam session and a weekend brunch with Mitch Resnick. When I pointed out that most people might not list brunch with their thesis advisor as "fun," he laughed and pointed out that *"the things I do for fun cross over with the things I do in lab. [There's] not a very good separation."* I'd counter that rather than this being a poor separation, it is an example of a wonderful fusion.

Personally, I've built my academic career on the importance of play. A few years into a tenure track faculty position teaching engineering, I realized that as much as I cared about my research topic at the time (projects related to design for aging), it was through PK–12 education and, in particular, technological literacy that I believed I personally could make the biggest difference in the world. So I traded in my lab full of walker parts and schedule of nursing home visits for, quite literally, paint, clay, and preschoolers. In the process of teaching college-level engineering classes, I was seeing a lot of basic misunderstandings about how electricity and circuits work. At the time, my daughter was nearing her first birthday, and I tried to imagine how I might teach her about circuitry. While makers like Leah Buechley were doing amazing work with sewn and painted circuits, I wanted something even easier to work with. Thus, I challenged a first-year engineering student, Samuel Johnson, to work with me for a summer to develop sculptable circuits. After diving into established scientific literature such as "1,001 Things to Do with your Toddler on a Rainy Day," we managed to fill our lab space with a brightly hued variety of play doughs, covering a wide range of textures, smells, and varying levels of electrical resistance. By the end of the summer, Samuel and I had developed recipes for conductive and nonconductive, nontoxic play dough, and Squishy Circuits were born. These days, we're delighted to note that Squishy Circuits are used in schools and museums around the world. Our goal, which seems to have been met, was to invite children of all ages to *play* with circuits (Figure 3-7).

Figure 3-7. A young child building a Squishy Circuit using conductive and insulating dough

I've found that very few people, when faced with a lump of play dough and LEDs, can resist playing with them. The feel of the dough, the lights and buzzers, and the scent of fresh play dough hold such a sway over people, that it has become an unofficial test for potential collaborators in our lab. Whenever someone is interested in the work that my students and I do in our Playful Learning group, I try to arrange for them to come play with some Squishy Circuits. More often than not, this leads to some laughter and a discussion about learning and possible directions we could take our research. Once in a while, though, I get a different response, a reply along the lines of "Why would you do this? What's the point?" Unsurprisingly, none of them have become collaborators of ours. My theory being that if the idea of building play dough circuits can't bring a smile to your face, you're probably not going to be too much fun to collaborate with. I've found that my best researchers and collaborators are also my best playmates.

As we'll see in Chapter 4, many young makers became fascinated with things that explode, ignite, or zoom around. In none of these cases were the children trying to cause trouble; rather, they were following their curiosity and reveling in the process. They were playing. And, often, their youthful pastimes gave hints of the makers that they would grow up to be.

To me, there is no wiggle room here. Play is essential to human development. Not just a nicety that we can take part in "if there's time." Play doesn't need to be expensive. We don't need to give our children expensive gadgets or fancy sports equipment, but we do need to give them time and encouragement. Reflecting on the definition of play, we can also see how it relates to Chapter 2's exploration of curiosity. Play is allowing ourselves to choose activities that we find enjoyment in, without worrying about the outcomes. This isn't something we can do for our children. Rather, we are at our most supportive as parents and educators when we stand back and let our children run free in the direction of their choosing, whether that means sprinting through fields or compiling code.

4

Risk

Makers are willing to take on risk.

They aren't afraid to try things that haven't been done before.

Explosions came up often in makers' descriptions of their childhoods. I heard stories about things that were exploded, lighted, or blasted in a wide variety of ways. Looking back on this as adults, many of those who were young fire and explosives enthusiasts admit that they are shocked that they were allowed to do the things that they did. One interviewee, who had a long list of fiery exploits in his childhood, told me that *"I wanted to talk to [my parents] before talking to you to find out what the hell they were thinking."*

I will admit that I had a hard time deciding whether to share these stories as I found myself realizing how close many of my friends had come to killing themselves. One friend recalled that he was only spanked twice as a child, and one of those times came at the age of four when his mother found a pile of discarded matches in the basement. They were leftovers from the *"controlled burns"* that he enjoyed doing, solo and in secret, as a preschooler. Almost uniformly, these makers expressed their hopes that their own children wouldn't try the things that they themselves had done. That said, a fascination with fire and explosives was an oft-repeated theme among those I interviewed for this book, and I doubt that that is a coincidence.

Hearing makers reflect, as adults, on their youthful experimentation was thought provoking. I have to wonder whether these adult makers would be doing the things they do now, had they not had these experiences as children.

One maker, now a father and an accomplished engineer, wrote:

> *At my job now, I work with some very hazardous chemicals, as well as machine tools, electricity, and extreme temperatures. I've been injured (at home, not at work), luckily not badly, and seen others injured. I have a much different sense of safety now and what is an acceptable risk to take on. I shudder, and my blood runs cold when I think about the things I did when I was young and the thousand ways things could have gone very, very wrong. I have few regrets in life, but if I could go back in time and refocus my interests somehow when I was a teenager, I would do it. It is only through sheer, dumb luck that, to my knowledge, no one ever came to any harm through my callous disregard for risk.*

Rockets to Robots

It turns out that even the person who taught me how to be safe in a machine shop had a history of youthful risk taking. When I started thinking about people to interview, Woodie Flowers (Figure 4-1) was at the top of my list. This self-described "weird kid" from Louisiana went on to become an Eagle Scout, an acclaimed MIT professor, and the co-founder of the FIRST Robotics Competition. I was fortunate enough to have him as my lab instructor for the much-celebrated 2.007 Introduction to Design and Manufacturing course at MIT. Despite being an ocean engineering major, I signed up for this Mechanical Engineering course because I desperately wanted to be the sort of person who could have an idea and then actually build it. I lost a lot of sleep over that class as a 20-year-old. I saw each machine tool as another way I could chop off an appendage or, at the very least, finally prove that I wasn't MIT material. Woodie always seemed to have time for me, whether it meant riding his unicycle down the hallway at 7:30 a.m. to go over my design, or to patiently show me, yet again, how to use the milling machine.

Coming from a family that spent considerable time fishing, hunting, camping, and working on cars together, making and fixing things have always been part of Woodie's life (Figure 4-2). As a teenager, Woodie and his friends were obsessed with rockets. Serious rockets. The fuel that they mixed contained lead and sulfur. They were also able to get lead paint from a friend who could procure it at the oil field, and at that time cans of industrial lead paint came with an extra container of powdered lead to be mixed into the paint. (This was before lead paint was outlawed due to its toxicity.) Woodie had a workshop in the back of the house where the boys built their rockets. Once, during a demonstration, a two-and-a-half-foot-tall rocket fell over, skittered across the ground, and "almost took out" the group of Boy Scouts who were watching. Sometime after that event, Woodie was called to the principal's office. This was the only visit he'd made to the principal in high school, and when he got there he was advised he should stop building the rockets because they were dangerous. Woodie admits that this was good advice, and he considers himself lucky that neither he nor others were killed by his rocketry experiments.

Figure 4-1. Woodie and his family camping deep in the swamps of Louisiana on a family hunting trip (photo courtesy of Woodie Flowers)

Woodie's father was the mentor who taught him how to build things well, and safely. He credits his father for teaching him more about engineering than his undergraduate engineering career. Lots of Woodie's spare time growing up was spent in his father's welding shop, where he learned to take care of tools and respect machines. Because their family had very little money, buying Woodie a car in high school wasn't possible. Instead, an uncle gave him a 1947 sedan that was so worn out that the wheels leaned over to the side. Woodie decided that he wanted to turn it into a hot rod. His father's response was *"OK, Scooter. I'll help you, but if you start it you have to finish it."* Both men stuck by their commitments and worked together to build a V8-engine-powered "hillbilly hot rod" that became the fastest accelerating car in town. Woodie's youthful curiosity spanned a wide range of topics. When he wasn't working on his car or rockets, Woodie was off collecting butterflies, which led to an award-winning science fair project on the impact of environment on Lepidoptera (the order of insects that includes butterflies).

Figure 4-2. A young Woodie with an injured thumb, a condition that he says "was typical of me as a kid" (photo courtesy of Woodie Flowers)

Just as his father had taught him how to build things, Woodie has been teaching these skills to others throughout his entire career. As a Ph.D. student in engineering at MIT, he minored in art at the Museum School and eventually became a teaching assistant in the architecture school because of his ability to turn his ideas into tangible objects. He firmly believes in the importance of hands-on learning, but also recognizes why that is sometimes avoided. *"You have to learn things by bumping into things. And that's scary for a parent, I'm sure. Simulations are good to a point."* Woodie taught MIT's Introduction to Design and Manufacturing course, in which students spend a semester individually building a machine to compete in an end-of-semester competition, and turned it into one of the campus's most anticipated events. However, whereas some competitions seem to bring out the worst in people, Woodie works to instill *"gracious professionalism"* in all of his students. (He defines that term as *"Compete like crazy, but treat each other well."*)

These days, Woodie inspires thousands of kids around the world to design and build robots through the FIRST Robotics Competition. Each year more than 300,000 students in more than 60 countries spend six weeks building machines to participate in a competition that requires not only excellent engineering, but also exceptional gracious professionalism to win. Having been a judge for FIRST, I often get to walk through the competition pits where teams, all clad in safety glasses and helping each other, are excitedly working on their robots. These are teenagers, many of whom are beginners, confidently using potentially dangerous equipment to finish robots that often weigh more than 100 pounds. There isn't a single runaway rocket to be seen, and participants can enter animations into a special competition for safety videos.

Gaining Competencies

When I sat down to talk to Christy Canida, she showed me pictures from a recent duck hunting trip that she'd been on. Her daughter, Corvidae (Figure 4-3), had been incredibly excited to join her on this outing. Mother and daughter dressed in hunting gear and met their group in the chilly, early hours of the morning. The outing, alas, wasn't without tears. It turns out that Corvidae hadn't realized she would not be allowed to help with the actual shooting. The disappointment is understandable because Corvidae is a skilled outdoorswoman, and an active member of the Bay Area Trackers survival skill group, where she's learned how to light a fire, whittle, and orienteer. She has been making the family's scrambled eggs solo for years now, and is well involved in her family's day-to-day activities. Unlike many other four-year-olds, Corvidae is well aware of where her food comes from, and is an active participant in its preparation.

For Christy and her husband Eric, the overall goal of their parenting strategy is for their children to know their limits and gain useful skills. Christy wants her children to *"be generally competent: I want them to have a strong set of basic skills, be able to pick up any new skill they're interested in, and understand how to do dangerous things safely."*

Figure 4-3. Corvidae holding the duck that her mother shot on their first duck hunting excursion (photo courtesy of Christy Canida)

Part of her philosophy is that it's important to allow small injuries in childhood so that their children can learn their limits and avoid the big injuries later. For this reason, she will encourage them to jump off of things of various heights so that she can teach them how to practice having "soft feet" when they land, and judging how high they can do this from safely. A conversation that is often repeated in their household goes something like this:

Child: *"Mom, can I do <insert reasonable, yet risk-bearing, activity here>?"*

Parent: *"What am I going to tell you if it hurts?"*

Child: *"Suck it up and walk it off."*

Parent: *"OK. It's your decision."*

We live at a time, and many of us in a culture, where people often have very vocal opinions about what it means for children to be safe. As you can imagine, the "suck it up and walk it off" approach counters many parents' beliefs on how to raise kids today. Christy describes being criticized by strangers. One day, recently, on a trip to a children's museum, Corvidae was climbing on one of the play areas and tripped, causing her to sniffle a bit. Christy just watched, patiently and quietly, waiting to allow Corvidae to assess the situation herself. A woman observed this, and Christy's reaction, and ran over to pick up Corvidae, while murmuring *"poor baby... are you OK?"* and glaring at Christy. The result was that Corvidae thought the woman was overreacting, and was more alarmed by this stranger picking her up than she was by the minor bruise she might get. Corvidae and her brother have grown up in a household where they are expected to participate in the day-to-day activities of the family as deemed appropriate by their parents.

Christy grew up in a small Midwestern town which, she says, necessitated the development of independence and self-sufficiency. If she needed something special, she probably had to make it, because she couldn't *"just order it from Amazon."* Her grandfather was a math major who grew up during the Depression and worked his way through a degree at Berea College. Christy remembers him as someone who could make and fix anything, and she and her brother were accordingly drawn to him and his shop. Hand in hand with this self-sufficiency went responsibility. He introduced them to tools and techniques for making things, and stressed the importance of respecting their tools. Christy can still recite his advice to them: *"If you have tools, you always clean your tools."*

When I asked Christy about her own parents' threshold for their children's risk taking, she happily told me about fireworks. Because many of the adults in her family enjoyed fireworks and rockets, the children were shown how to safely participate. She remembers being *"single-digit aged"* and setting off bottle rockets with her uncles. She still clearly remembers one time when the police came to investigate the noise from the explosions. Upon seeing the police cars, her adult uncles hid, leaving her and her brother sitting with the bottle rockets and the lighters. The officers saw the children and the fireworks, asked the kids to make sure they angled the rockets away from any tents in the area, and then left. Apparently they were not overly concerned about two small children launching rockets. In all of her stories it was made clear that her family stressed to her the importance of respecting tools and dangers, and of learning (and appreciating) their own limits.

As we discussed the childhood activities of herself and her children, Christy pointed out the disparity between what today's children, particularly in the United States, are allowed and expected to do, and what children multiple generations ago showed themselves capable of. Some may argue that this kind of behavior could lead to injury. They're right. Christy was quick to show me a large number of scars on her hands, but noted with pride that she hadn't broken any bones until she reached her 30s. Having been allowed to get small injuries seems to have helped prevent the larger ones.

Christy's adventurous childhood led to a degree in biology from MIT, and a day job that involves enabling, and encouraging, people to teach each other how to make things as the Community and Marketing Director for Instructables. She is a prolific maker herself, having posted instructions for more than 170 projects ranging from handmade baby shoes to squid ink harvesting, taxidermy to woodworking, and a wide array of creative Halloween costumes.

What Is Safe?

Once, after giving a talk about the importance of making in childhood, I was approached by a preschool teacher who told me that in her class of three- and four-year-olds, she was only allowed to let the kids use pencils if there would be one-on-one supervision. Other educators speak of how sewing could no longer be done in their school because children could prick their fingers (and at least one person mentioned that sewing involved "sharing needles," which could send the wrong message to the children). Many educators have come up with fantastic workarounds to rules like this, such as the museum staff member who taught sewing with "non-pokey" plastic needles pulling yarn through loosely woven sweater fabric. I have to wonder, though, whether there comes a point where we're actually being negligent by overprotecting our children from risk. Christy's suggestion that we allow children to have small injuries in hopes of preventing later major injuries seems reasonable. Opportunities for children to learn through these small injuries, however, seem increasingly harder to find. Particularly when it comes to playtime.

I run into this quandary often as I look for tools to buy my own children. When my oldest daughter was nearly four years old, I decided it was time to get her some hand tools. I looked online for construction sets geared toward children. Thinking fondly of the sets I had when I was little, I looked closely to see if I could find one suited for my kids. I was intrigued by one kit that promised "real" construction play. While the kits that I played with in elementary school typically included glue, nails, and a rough picture of something I could build with a hammer and maybe a saw, this kit included foam "wood," plastic tools, and plastic nails. Even more surprising, this toy is labeled for children ages 6+ and had a manufacturer's recommended age range of 6–15 on Amazon.com. Minutes earlier I'd been confidently pricing hand drills and hammers. Now this toy seemed to be telling me I should wait on those tools until my daughter reaches middle school. So how old is old enough to hand kids real tools?

The more I research children and tool use, the more I notice how dramatically things have changed. Kids were once trusted with real, metal tools. In the early 20th century, it was common for elementary schools to teach manual training. In 1900, Frank Ball, a teacher at the University Elementary School in Chicago, wrote, *"At the present time no thoroughly equipped school is complete without its department of manual training or construction work."*[1] A book written in 1964 by John Feirer and John Lindbeck, of the Industrial Education Department at Western Michigan University, talks about outfitting elementary school shops and advises that the tools should be maintained well, because *"the sharp, well-cared-for tool is safe, easy and fun to use."*[2] Very rarely these days do we hear "fun," "sharp," and "elementary school" in the same conversation.

Making objects is similar to making music. We would think it outrageous to wait until students reach university to give them their first non-toy musical instrument. However, many students reach their first year of college without much experience with tools. An engineering professor told me he asked a class of 35 first-year engineering students how many had used a drill press. Not a single hand went up. He continued, asking how many had taken apart one of their toys when they were younger. Again, not a single student raised a hand. This was a roomful of future engineers.

Had Lenore Edman (Figure 4-4) gone into engineering (she majored in interdisciplinary studies with an emphasis on English and Greek), she would have been able to raise her hand. Lenore, and her husband Windell Oskay, are the founders of Evil Mad Scientist Laboratories, a company that designs, manufactures, and sells DIY kits. She has worked on projects ranging from a wooden mechanical computer that uses ball bearings to compute mathematic operations, to a robot that turns iPad doodles into watercolor paintings, and on to furniture that senses, and reacts to, the user's movements.

1. John Dewey, *The Elementary School Record* (Chicago: University of Chicago Press, 1900.)
2. Feirer, John Louis, and John Robert Lindbeck. *Industrial Arts Education* (Washington: The Center for Applied Research in Education, 1964.)

Figure 4-4. Lenore holding a picture of herself in first grade—if you look closely, you'll notice that she is wearing the same earrings in both pictures (photo courtesy of Lenore Edman)

Lenore was still in elementary school when she assisted her father in designing and building a three-story-high tree house (Figure 4-5) using found materials. After she gave me an elaborate description of the tree house's features, I asked at what age her father had trusted her with real tools. Lenore found the question impossible to answer, because she couldn't remember a time *before* being allowed to use them. So encouraged was she to learn how to build things, that her dad always left a block of wood, a bucket of nails, and a hammer out so that anyone who wanted to could start hammering away. She truly meant everyone; even her two-year-old brother was given access to the bucket of nails and hammer.

Figure 4-5. The three-story-high tree house that Lenore's family built using found materials (photo courtesy of Marlo Edman)

Chemistry Sets

For kids who were interested in explosions and fire in the 1960s, there was *The Golden Book of Chemistry Experiments*. This book had a note to adults stating that they should supervise youth who are using it and, in the process, will probably learn something themselves. *The Golden Book of Chemistry Experiments* by Robert Brent and Harry Lazarus (Golden Press) taught kids how to do things like make chlorine gas. The book mentions potential dangers and is very careful, but the authors and publisher of this book trust that a child with the right mentor and the right supervision can be trusted with experiments like this. In contrast, consider one of the chemistry sets available for kids today. On the front of the box it highlights that kids can use it for fun chemistry activities. Also on the front, it reassures parents that this set does not contain any chemicals.

The 1960s was also the era that brought us the Atomic Energy Lab kit for kids. Jeffrey Jalkio, a fellow engineering professor at the University of St. Thomas, had both this and *The Golden Book of Chemistry Experiments* as a child. When he heard that I was researching maker childhoods he started bringing in some of his old toys to show me. While I had thought that *The Golden Book of Chemistry Experiments* was the benchmark for trusting children with dangerous material, the instruction manual for the radiation kit quickly topped that. Jeff did note, somewhat wryly, that he hadn't known that he owned this kit as a child. It wasn't until recently that he found it in his mother's attic. Apparently even a mother who let her son play with many interesting toys, had concerns about handing her son radioactive materials.

Trusting My Own Children

After speaking to so many makers who shared their adventures with real tools and real materials, I was even more determined than ever to introduce my daughters to woodworking. This ended up being surprisingly difficult. I started by going to a local "big box" chain hardware/home-repair store. When I got there, I asked a sales clerk in the tools section for suggestions. I couldn't help but notice that he looked uncomfortable when I said I was shopping for a child. He waffled a bit about how they didn't really sell things for children, and then he asked how old my daughter is. Here I must confess that I lied and said eight. (I figured that if they couldn't help me shop for an eight-year-old, a four-year-old was out of the question.) The reply? He looked at me a bit wide eyed and said. *"Eight? That's really young. I didn't have woodshop until my last year of middle school, and that school doesn't even do that anymore. Eight?"* This is when I knew that this shopping excursion was going to be a challenge, and decided to leave.

Later that day, I was in a small shopping district buying some other presents. As I walked to my car I noticed a small hardware store on the corner. On a whim, I went in, not expecting much. However, when I said that I was trying to buy tools for my daughter, two of the sales clerks jumped up. The first suggested I look at the smaller hammer that they had just sold to a dad and son, the other started walking me down the aisles pointing things out. Then they asked "the question." *"How old is your daughter?"* When I said four, they nodded and said that was a bit young, but that was all. I left with a selection of tools for my daughter, who has since gone on to build small projects like doll furniture. This corner hardware store was the kind of place where people who have been using tools their entire lives were eager to spark that interest in a new generation. This was the kind of store that could become a treasure trove of materials for my daughters as they grew up. Unfortunately, this is also the kind of shop that doesn't do well against the big-box chains. When I next returned to that store I found that it was no longer a hardware store, but a place offering photo sessions for pets.

I'm not going to ask Jeff to loan my daughters his Atomic Energy Lab set any time soon (or ever), and I'll confess that I'm secretly hoping that explosions don't make it onto their list of interests. But discussions with makers like Christy, Lenore, and Woodie have definitely caused me to rethink the risks I allow my daughters to accept. My younger daughter has been sewing with a real needle since she was two, and my six-year-old can confidently use a sewing machine by herself. The training wheels are coming off the oldest's bike, and the three-year-old is starting to take solo walks to a neighbor's house to borrow books from their "Little Free Library." There are a lot of potentially dangerous things in the world, but I'm also coming to realize that obsessively avoiding all possible risks is neither feasible, nor desirable. I should probably take some comfort in my daughter's new tendency to always keep some bandages in her backpack.

Acceptable Risk

How much risk is acceptable? How do we teach makers, young and old, how to assess risk? Many people talk about trusting their instincts and I think this was put most eloquently by restaurateur Nick Kokonas. Quoting advice that a friend had given him about business, he says *"if you feel fine when you go to sleep and wake up, you are not taking enough risk. If you are throwing up constantly, you are taking too much risk. A nice, even nausea most of the time is just right."* Nick is a father, and a maker, so I asked him if he would apply this advice to parenting as well. He admitted that while he wants his kids to feel that way about academics, business, and art, he's "probably overprotective" as a parent. I took this as a sign that he himself hadn't been a child who took many physical risks. I couldn't have been farther from the truth. Yet again, I got to hear a story about a kid exploding things in his backyard:

I once emptied out the black gun powder of several of the D size Estes model rocket engines. I wrapped it in paper towels, put it in a Styrofoam cup, and lit it on fire. The ensuing white-smoke mushroom cloud was impressive, but I had hoped for an explosion. I read all I could find in the library on making gunpowder, "bombs" etc. Nowadays they'd have me in jail for sure thinking I wanted to do harm... but of course I just liked blowing stuff up.

Anyways, I eventually found out somehow that compression was important to explosions. So enter: duct tape. I bought more model rocket engines at the local "dime store," emptied them out, ground the gunpowder fine (somehow I knew that would burn more quickly), put it into an empty aspirin bottle, drilled a hole in the cap, put a long wick strung together from firecrackers in it, and wrapped an entire roll of duct tape around it until it looked like a cannonball, dug a small hole in the backyard, put it half in, half out... lit and walked about 10 feet away. I didn't know what I was expecting but the ensuing explosion was very, very real. One of my neighbor's windows cracked (at least 50 yards away). I had ringing ears. And the fire department was there very quickly. My dad came home to find the fire trucks. I concocted a story about a model rocket gone wrong, extra powder, etc. The fire department was fine. My dad knew better. I do NOT tell my kids that story!

Nick was unsure about whether he'd let his own kids blow thing up, something many makers admit they're hesitant to let their own children do. *"It's a very different world now... Back then I was considered clever—now I'm sure that I'd be seen as some sort of threat."* While Nick decided he likely would let his kids dabble in explosions, he also said *"I don't want it on YouTube."* The pause before he decided that he'd let his kids try things like this is telling.

Every maker who told me stories about explosions in their childhood voiced concern over whether they could, or should, let their own children do similar things. Others debated at what age they would let their children use various tools. As the definition of "safe" with regards to children's behavior and activities evolves, the question is raised as to whether we can teach children to be innovators if we don't let them learn how to assess risk. It's our task to help the young makers in our lives, as Christy Canida put it, *"understand how to do dangerous things safely."* After all, any new endeavor and project, be it building a boat or crossing a street, brings risk.

5

Responsibility

Makers take on responsibility.

They enjoy taking on projects that can help others.

When I started teaching in Minnesota, I heard a common refrain among older engineering professors: *"The farm kids are gone. Get over it."* Eventually, I learned enough about Midwest engineering culture to understand what they were trying to convey. Traditionally, many of our strong engineering students came from farming backgrounds. They would arrive at the university with hands-on experience maintaining and building equipment. A senior executive at a Fortune 500 company in Minnesota once told me that his "dream hire" for technical positions is an individual with a Ph.D. in a STEM discipline who also spent his or her childhood on a farm. The number of job applicants fitting those criteria is small and dwindling. I would propose to this company (and others) that they start looking instead for new hires who are lifelong makers.

Farm Kids

While the mechanical savvy that many "farm kids" possess is often discussed, I see that as just one attribute shared among this group. Farm families depend on all members to do their part in getting the work done, and thus most farm kids grow up with a strong sense of responsibility. Steve Hoefer (Figure 5-1), who grew up on his family's farm, is now a designer who is often hired to help solve technical problems for companies while also creating a series of how-to and DIY videos. His farming background is invaluable not only because he had freedom and access to real tools, but because all family members, regardless of age, were expected to pull their weight and participate in farm life. Even the smallest kid can help on a farm, and often has to. On a farm, Steve explained, if you see a loose bolt, you start turning it. It's just expected. The family depends on the farm for their livelihood, and thus it is of crucial importance that everyone around pitches in and gets things done.

Steve said that farming also instilled in him respect for the talents, especially the unexpected talents, of the people he works with (Figure 5-2).

Figure 5-1. An 11-year-old Steve Hoefer gardening (photo courtesy of Jean Hoefer)

As an adult, Steve prides himself on his comfort in new situations and his confidence to get work done even in the face of uncertainty and doubt. It is this ability, coupled with strong technical abilities, that has led companies such as Sony, Microsoft, and Leapfrog to seek his assistance when designing prototypes to explore new product ideas. Steve is the kind of innovator who can develop cost-effective, and novel, solutions to problems while always bringing a sense of whimsy to his work. Steve maintains that a childhood on a farm instilled this in him. So many of his daily tasks as a child could be summed up as *"Go and do something you've never done before. Figure it out. Learn something. Maybe even discover a better way of doing it."* Steve recalled, *"[T]here were daily events where we were told to go off and do something, usually important, given the tools and materials, and the rest we had to figure out for ourselves. And usually it worked out. And when it didn't, it wasn't the end of the world."*

When I asked him for an example of being given responsibility as a child he said that there were so many stories he could tell, but that the one that stood out was also one of his earliest memories. When Steve was about five years old, his family was driving cattle between two pastures that were half a mile from each other. Steve was still small enough that a *"cow had to look down to see the top of [his] head."* That didn't deter Steve's dad from assigning him a role in the cattle drive. Steve was given a stick and told to stand at the destination pasture's gate. When the cows arrived it would be his job to guide them through the gate.

Once Steve had been given his instructions, the adults left him so that they could be behind the cattle, steering them in the correct direction. Steve still vividly recalls that day:

> For at least half an hour I'm standing there by myself, sweating with anticipation. I was small enough a cow could knock me down and not notice. The stick didn't make me feel especially powerful, and if I failed to turn the cows they would just keep going down the road and into the neighbor's field, a total disaster. But my dad knew our livestock and understood that when they saw the open gate and the fresh pasture beyond they'd be glad to go in. All I had to do was make enough noise to get them to notice. When the cows finally came trotting over the hill towards me I grabbed my stick in my tiny little fist and started yelling and waving. That was all it took. The lead cow looked around, saw the greener pastures and headed in, leading the rest.

Imagine, for a moment, the power that a child feels upon successfully guiding a herd of cows to safety using nothing but a stick. Maybe even more important, imagine how it feels to know that your father trusts you with this task. Steve recalls that sense of accomplishment vividly, maintaining that *"afterward I knew that a little boy with a stick could turn an entire herd of cows."* Children who have been given the opportunity to try something important—to succeed at something important— tend to grow up with ideas for all sorts of amazing and impactful tasks they can complete as they get older. If a five-year-old can herd cows, what can a ten-year-old accomplish? Or a twenty-year-old? That sense of possibility is clearly the message Steve's upbringing gave to him.

I heard similar sentiments from others, now grown and far from the fields of their childhoods, about how the lessons they learned growing up on a farm have influenced their current endeavors. Teamwork is vital. A willingness to pitch in and do what needs to get done played a part in almost all of their stories. Given that most of the makers I spoke to live in cities, I was actually surprised at how many of them had spent at least part of their childhoods on family farms. In many cases, this involved visiting grandparents or relatives. Even short or occasional visits to those environments seem to have made disproportionately large impacts.

Figure 5-2. Steve Hoefer, at four years old, being taught about ants by his sister Jennifer (photo courtesy of Jean Hoefer)

Butter Knives and VCRs

For those of us raising kids far from fields and barns, it's reassuring to realize that kids don't need to live on a farm to learn how to take on responsibility. Luz Rivas (Figure 5-3), engineer and founder of DIY Girls, grew up in Los Angeles with her mother and sister. From an early age, Luz was the one her family would turn to when things broke or needed assembly. She notes that, culturally, this role was a bit unusual. *"If there's a man in the house, they fix everything. In my house there wasn't a man,"* so it was Luz who stepped up and took care of the building, fixing, and repairing. *"Anytime something needed to be fixed or built, they'd call on me... Even as a kid, being six years old, it would be 'Luz we just got this, you need to figure out how to connect all of this.'"* This is how it came to be that the Rivas family got their TV antenna fixed, their VCR set up, and even, once, a functioning Atari installed despite its not coming with all of the necessary connectors. Setting up the Atari took the eight-year-old Luz multiple hours. She persisted, though, because she saw this as her job in the family.

Luz took her role seriously, especially when her tasks were difficult. Even from a very young age, her determination to contribute instilled in her a sense of desire and, eventually, of accomplishment. When it came time for Luz to leave home for college, her mother and sister would call her up and have her troubleshoot their electronics over the phone.

Figure 5-3. A young Luz Rivas (photo courtesy of Luz Rivas)

One of the early steps in most projects is identifying what tools will be necessary. Luz's family didn't own many tools, so she improvised. Lacking a Phillips head screwdriver, she dedicated one of their butter knives to this role, telling her mother and sister that *"this is a tool. We're not going to use it to spread butter or anything."* Later, she added a small steak knife to her toolbox. Telling me these stories, she starts laughing, *"Oh my god. Why was I using the utensils? I would walk around the house and find something that works."*

Luz identifies that kind of creativity and improvisation as a sign of a maker. To her, a maker is *"anybody that has an idea and uses the tools around them, that they have available, to create it."* Growing up in Los Angeles surrounded by Mexican immigrants, both as a child and now, Luz has learned that communities such as hers are full of makers. Being part of the community means learning who can be turned to for different needs. Some people can *"make anything out of metal"* for you, and others can create clothing. Luz's grandmother fit this latter category. When

Luz speaks about her grandmother, the pride is palpable despite their never having met. Luz grew up listening to her mother tell stories about how Luz's grandmother was known throughout their community in Mexico as a woman who could simply look at someone and create, without a pattern or any formal training, a fabulous dress. Through her dressmaking she was able to support her family, and also become a valued member of her community.

Luz's commitment to family, community, and the empowerment of girls is inspiring. She has served as an assistant director at Caltech managing outreach programs for underrepresented students in science and engineering. She has advised chapters of the Society of Hispanic Professional Engineers. She has developed engineering programs in technology for children and families. These are each major undertakings benefiting a variety of makers, and they only represent a small portion of the work that Luz has done.

Three years ago, she took what she had learned through all of these programs and founded DIY Girls. DIY Girls is a nonprofit organization in Los Angeles whose mission is to increase women's and girls' interest in technology, engineering, and making by providing hands-on educational experiences. Anyone who follows Luz's Twitter feed is used to seeing pictures of young girls intensely focused on their soldering irons, or working on a piece of code. She is particularly focused on helping elementary school girls gain confidence in fields that have historically been less welcoming to them. *"I want [the girls] to leave elementary school and enter middle school more confident about their own abilities whether it's technical [or] making. They go into middle school knowing that they are able to do a lot of this."*

Luz, like Eric Rosenbaum (see Chapter 3), cautions against making everything a competition for kids. Given the constraints and rules of most competitions, the time pressure often makes it so that the faster kids on a team take on most of the work and the kids who need more time fall by the wayside. Years ago, Luz served as a coach for a youth robotics team. At one competition, the robot built by Luz's team broke. One of the girls on the team quickly rushed out on the field to fix it. The team, and coaches, began screaming *"leave it!"* because touching the machine would cost them points. This led to the girl crying, likely out of embarrassment for almost costing her team points. Luz was mystified by this rule, asking, *"What message are we sending that something breaks down and your first instinct is to fix it and we're yelling no because we'll lose points? Who cares about those points?"* Even without losing those points, their robot didn't do particularly well, and there were many tears spilled by the young girls on the team. After this incident, Luz realized that her passion was in getting the children to love learning for the sake of learning, as opposed to focusing on a competition. She notes with pride that many of the girls in DIY Girls ask her to take pictures of them wearing their safety glasses and working on projects, and then to send the photos to their dads! Just like a young Luz, these girls are proud of their ability to do useful things that can make a difference for their family and community.

Big Wheels and Code

Amon Millner (Figure 5-4) is another maker who learned, from a young age, the power of being able to build and fix things for one's family. On Christmas morning when Amon was five, he was excited to find a shiny new Big Wheel under his family's Christmas tree. Unfortunately, as soon as Amon jumped on it, he got his first lesson in mechanical failure. Due to an improper configuration of parts, the pedals couldn't turn. Driven by his desire to speed around the neighborhood, the pint-sized Amon began investigating the vehicle and discovered a nut that wasn't screwed in.

When Amon relates this story to me, it's clear that rather than being disappointed that his Big Wheel didn't work right away, he was empowered by realizing that he, as a five-year-old, could fix something and make it usable. This was also a turning point in his relationship with his father, who had been the original, and unsuccessful, Big Wheel mechanic. Amon recalls with appreciation that his father put him in a position to be a maker that day, and that that was a strong theme in his life going forward. With parents who were skilled educators (but not particularly skilled at or interested in building tangible things), he had local support for following his interests, but looked beyond his household for making-related mentorship. He cherished the time that he spent shadowing his maternal grandfather, who would undertake projects to improve Amon's family's house every time he would visit from his home 3,000 miles away. Although Amon would only have a short window overlapping with his maker grandfather, their time together left him with some idea of how to use the drawer full of tools that he left behind, bequeathed to his tiny apprentice.

When things didn't work in his household, young Amon became the one who would step up to fix them. Or, in many cases, he was the first one volunteered for this role by other family members. Supporting Amon's belief that *"everybody is [born] a maker and some get to stay that way longer,"* he recalls the environment that he was raised in as one where he was encouraged to tinker, fix, and make. This role took on added importance because he considers his parents superheroes for being able to provide him with everything he needed, but he was well aware of the resourcefulness had needed to employ to make that possible—and the value of maintaining possessions when replacing certain items would put a strain on available financial resources. Thus, Amon's ability to repair devices was valued highly by his family. Broken things also supplied a low-risk opportunity for the young maker. They were already broken, so the worst that could happen after Amon tried to fix them was that they would remain broken and end up in the trash. On the other hand, more than occasionally, Amon was able to return the devices to a usable condition, which both he and his family considered an almost magical occurrence.

Figure 5-4. Amon Millner and his sister playing in a glass cabinet that they pretended was a TV screen (photo courtesy of Amon Milner)

Middle school marked an important turning point for Amon, because he was introduced to programming computers for the first time. He considers himself to have been at the right place at the right time, because he attended a school that received infrastructure intended to address some of the disparities between resources available to affluent suburban schools and urban schools in his city. That meant a computer lab with a dedicated instructor. Throughout his middle school years (and beyond), Amon took every programming-related offering that his computer teacher Mr. Moore put on the books. He remembers this time fondly, recalling various games that he created and played.

Unfortunately, when Amon got to high school there was no longer a primary computer teacher, and a math teacher was assigned to teach the students in the computer classes. Unlike middle school, where he had an engaged teacher, his high school experience in the subject lacked someone who was passionate about guiding him through the rapidly changing field. This did, however, provide Amon with an opportunity to discover his love of teaching. If the teacher was absent, Amon was allowed to take over the class. Not so different from his days fixing his family's electronics, Amon was able to use his knowledge and skill to help others and be placed in a role of responsibility. When his school's guidance office needed someone to create a web page, they looked for seniors, but found no one available or interested in the task. They gave Amon the task, unaware that he was only a sophomore.

So impressed were the school administrators by Amon's website, that when a small software company across the street came looking for a potential intern, they recommended 15-year-old Amon. This is how Amon became the fourth employee at a start-up company before he was even old enough to drive. There was the *"software guy,"* the *"hardware guy,"* the company president, and Amon, the high school sophomore. Given the size of the company, this meant that Amon was immediately given responsibility and taught all aspects of the company. He learned how to lay cable infrastructure for networks and make programs that connected to commercial databases. Best of all, when the other three employees were out of the office, he got to pretend he was the boss. This experience was life changing for Amon because, by his own admission, he wasn't a model student. Ironically, he admits that the college he now teaches at would have had to make a special exception if it were to even consider high-school Amon as a candidate because he never took some of the courses required for admission—such as calculus. However, the fact that he was writing software used by major corporations made up for some of the weaker aspects of his transcript, and he went to the University of Southern California (USC) to study computer science.

Amon stresses the importance of the welcoming nature of the Maker Movement, and how we need to work to make it even more inclusive. As a child, Amon was all too aware of some of the interactions between authority figures and youth that contributed to what is currently a STEM workforce that does not leverage the talent available from the diverse (along ethnic and gender lines) pool of potential innovators. He can tell countless stories of path-altering encounters faced by him and his peers. Amon elected to go out of his neighborhood to a high school that was an environment where many Caucasian students opened up pathways to positive futures for themselves, but had not been a welcoming and positive pipeline for many African-American students. He had heard stories from the school such as an African-American alumnus's experience of being told by a teacher that *"black people shouldn't be scientists, and that they should focus on athletics."* Amon's group of close friends who were making the high school transition with him were told to *"go back across the river"* by unwelcoming students at their orientation. Ignoring the misguided advice, Amon found supportive colleagues in high school, including the colleagues at the software company who invited Amon into the tech world with open arms. To him, the Maker Movement acknowledges that it takes a village to raise a child, and he works to bring this message to communities where children often don't have opportunities to interact with technology in playful and meaningful ways.

Making It Yours

Some of the hardest classes to get into at Stanford are those taught at the d.school. The d.school, as the Hasso Plattner Institute of Design is referred to, is often held up as the gold standard for how interdisciplinary design thinking and creativity can be taught. In no small part, that is because one of its founders, David Kelley, is an educator known for his tireless work in bringing the concept and tools of design thinking to a wide variety of applications and venues. As a designer, he founded IDEO, the renowned product design firm that created the first mouse for Apple. David's work has been recognized through honors including induction into the National Academy of Engineering, a Sir Misha Black Medal for "distinguished contribution to design education," and the National Design Award in Product Design.

David grew up watching other family members build things, and then trying to do it himself. His grandfather was a machinist and his uncles worked in factories. When David turned 12, he received a brand new bicycle for his birthday. It was an expensive gift, and thus a big deal. He loved it. And then the very next day he sanded the finish off of it and painted it a new color.

In many families, this behavior would not go over well with parents, who might read such an action as draining the value of a gift, or even accuse their child of being irresponsible with his belongings. However, that was not the way David's family responded. Which, to be honest, is a bit surprising, given that his earlier attempt to disassemble and then reassemble the family's piano had only been successful with regards to the disassembly stage.

A new coat of paint wasn't the end of the modifications for the birthday bike. When the bike and David were a bit older, he decided to turn it into a tandem bike by removing one of its wheels and then bolting it to another bike. Unfortunately, the design didn't take flexing into account and never worked the way David had intended it to. Despite the less than successful outcome of this bike experiment, he still recalls the *"cleverness, excitement of the idea... and the specific knowledge of how to do it... and the capability of doing it. All those things just feel good."* This is exactly the feeling he tries to instill in his students at Stanford, and in his own daughter. A phrase that comes up during my discussion with him is *"bias towards action."* David wants to make his students confident enough in their ideas that they start trying them rather than just talking about them.

This "bias towards action" is a key attitude held by most of the makers I've met. I would argue that this bias is an element of assuming responsibility. By taking action, we are choosing to take on responsibility for the outcome. This is a powerful position to be in, particularly for children who are realizing for the first time that they can make and do things their own way.

Often when I give talks I ask the audience members to raise their hands if they can remember something they made as a child that they were proud of. Typically, every hand goes up. Then I ask whether that object was made by strictly following instructions or as part of a class where every student made identical projects, and I get fewer raised hands. For most people, the maker projects that they remember fondly are the ones where they were able to make something that was uniquely theirs. People tell me stories of lopsided pottery and semi-functional electronics projects, and their pride is evident. Following someone else's plans or directions, even when it results in a more professional-looking outcome, is less satisfying than making something from scratch.

Trust Us

For many children, a sense of responsibility can only develop when their families trust them to complete—or even attempt—a new project. Whether it meant being allowed to take apart broken household devices or taking sandpaper and paint to a new gift without being yelled at, makers reiterated that trust during their childhoods instilled in them a sense of responsibility and empowerment.

The Maker Movement is one that recognizes, and rewards, skill and persistence, regardless of age. I attended Maker Faire Bay Area, the "big Maker Faire," for the first time in 2012 and was particularly impressed by young makers (some as young as 10 or 11) who were designing their own projects or even, in one case, starting their own company. At an age where many kids aren't allowed to stay home alone, these young makers were being encouraged to reach out and find mentors and supporters who could help them take their ideas to the next level. I often tell people that the Maker Movement is "age agnostic." If the most knowledgeable person about a topic online, or at a local makerspace, isn't even old enough to drive, that rarely matters. I've watched children teaching adults how to solder and program, and the most rewarding part of the experience is the shared respect between the learners and their younger mentors.

As we saw in Chapter 4, risk and responsibility have a complicated relationship with one another. Makers, particularly those who have become parents, often look back on some of their youthful projects with concern, and a sense of gratitude that no one was injured. More than one of the people I interviewed have since called up their parents to ask why they were allowed to do some of the things that they did when they were younger. I strongly believe, though, that we can't learn responsibility without some level of risk. It doesn't take much persistence, or bravery, or even curiosity, to do something that you know will work. But those vital values are established when children are entrusted to try something new, when success isn't certain or even likely.

We learn responsibility when there is a very real chance that things won't go perfectly. The cows that five-year-old Steve Hoefer was guiding could have run off course, but despite that possibility, his parents trusted him to help with the herding. Having no training in electronics, Luz could easily have damaged her family's electronics rather than fixing them, but through her persistence and sense of pride in her work she was willing to put in long, sometimes frustrating hours until she got the devices in working order.

As educators and parents, one of the most valuable gifts we can give the children around us is our trust. I found myself reflecting on this this past Mother's Day when my daughters awoke me at the crack of dawn with breakfast in bed. My six-year-old was quite proud of herself because she had been allowed to make much of the food and carry a tray, complete with various cups of liquids, up the stairs to my room. My three-year-old followed her, beaming and carrying the milk for my cereal. I'll admit that my first thought was *"Crud. They could have fallen on the stairs and shattered the cups or made a huge mess."* Both girls, though, told me about every aspect of the breakfast that they had been involved in and repeatedly used the phrase "I did it myself." Their pride was evident and practically radiating from them.

Responsibility is learned through attempting difficult and important tasks. When we go out of our way to make things easy, or only set simple challenges and tasks for children, we aren't allowing them to see themselves as someone who can take on complicated and important roles. What many makers have in common is that they were allowed to take on challenges and roles from very young ages. Even when the tasks didn't work out the way they had hoped, the fact that they were *trusted* to do these things strongly influenced how they viewed themselves and the world around them. Not only does this result in a powerful sense of responsibility, but it results in adults who understand the value of trust, and teach it to the next generation.

6

Persistence

Makers are persistent.

They don't give up easily.

If you ever need to Skype Luc Mayrand, you'll notice that his profile image is a picture of himself as a smiling child, holding a model rocket. Though it's been decades since Luc was that young child, his ability to dream big and then, almost magically, bring those dreams to life has remained. As an executive creative director at Disney Imagineering, and as a core lead in the development of Shanghai Disneyland, Luc develops experiences for visitors, which means leading and collaborating with a wide assortment of experts, ranging from writers to choreographers to engineers and composers. Clearly a job that requires creativity and stellar problem solving skills.

When I asked Luc about his childhood, he quickly let me know that he was *"the luckiest kid on earth."* Luc grew up in the suburbs of Montreal at a time when he and his fellow local children had, what seemed to them, complete freedom. He lived two blocks from a public library. Given the small size of the one-story building, Luc learned the layout well, and spent hours perusing the books and looking for inspiration. *"By the time I was 10 I knew the ins and outs of the Dewey Decimal system. I found books about making models of cathedrals out of paper. I'd find things and try. I always liked to tinker with stuff."* Luc describes a childhood full of art classes and adventures. He is quick to point out though, that despite doing well in school, it was hard for him. *"I wasn't one of those geniuses that picked stuff up in no time. I had to work. I had to work diligently. I was always fairly good at math and language, but I could push myself to do anything. I could push myself to study. To stick to something. I would do it. If I wanted something I could get into it and do it."* This persistence is a trait that seems to apply to everything Luc, as a child and as an adult, decides to tackle.

Luc considers himself fortunate to have been born in Montreal at a time when the city served as a canvas for two major international events. The Universal Expo took place in 1967, and while he was too young to really remember what he saw from his stroller at the Expo itself, the remnants of the Expo shaped his childhood. All of the grounds that were developed, including an entire island and an expanded subway system, remained. A portion of the budget had gone to fund artists who created installations around the city, and Luc clearly recalled being 10 years old and traveling through the city by himself, by bus, visiting these sites. Three years later, Montreal hosted the Olympics. Luc would walk to the worksites where they were constructing the Olympic buildings, two miles from his house, and press his nose between the boards so that he could watch the buildings evolving. Luc's family couldn't afford tickets to the sporting events, but would visit the public areas and watch the procession of athletes. The swirl of languages and sights left an indelible mark in Luc's memories. Later, the Olympic facilities were opened to the children of Montreal and Luc learned to swim in Olympic pools, and to sail in the catch basin that had held "Man and his World" for the Expo and rowing for the Olympics. The 1976 Olympics left a strong social memory and archaeological imprint on the city and its inhabitants, something that he sees as lacking in some more recent Olympics. Luc was also very aware of the American space program, and his parents took him on a trip to Cape Canaveral. To this little boy who swam where Olympic athletes swam, and who traveled freely through a city where art was accessible and abundant, wonder could be found everywhere. *"I thought that the world was something to be made and anything was possible. I grew up thinking the world was for us to make and to make anything of it."*

As a young child, Luc could often be found making things out of paper or found materials. A major turning point came when he found a little box in his grandparents' basement. Opening it, he found a dozen assorted rusted Meccano set pieces, which he quickly turned into a makeshift car. He proudly showed the car to his family, and his parents noticed how interested he seemed to be. Shortly thereafter, on Christmas, they gave him his first complete Meccano set, and each year he received a bigger, more complicated set. Eventually this included the 4EL set that had electric parts. Suddenly Luc could barely keep up with all of the things that he wanted to make. He began a quest to gather as many Meccano pieces as he could. Each kit included a book that showed all of the other kits and Luc would spend hours poring over the possibilities. And then, when he was 11, he saw Meccano Set Number 10. Even 30 years later, Luc can describe this set down to its finest details, including its 1974 price tag: $620. Despite the overwhelming cost, Luc convinced his parents to take him into the city to a store that sold Meccano to inquire about the kit. When they got there they learned that the store didn't stock the kit; in fact, *no stores in Canada* stocked it. It had to be special ordered from England. The price tag was way beyond anything his parents could afford for their son. Nonetheless, in that moment, Luc decided he would have Meccano Set Number 10. Just as he pushed himself to study in school, despite finding the material challenging, he knew that there had to be a way to acquire this treasure chest of possibilities.

Luc got a paper route. Quickly realizing that he would never be able to earn $620 with a single route, he added more routes and took on winter delivery. A year and a half later, Luc had amassed three paper routes, a wagon with wheels for the summer and skis for the winter, and a bank account with $320 in it. Not content to simply wait and grow his bank account, he also made calls to all of the hobby shops in Montreal to see if any had the kit. Not surprisingly, none did. However, one shop owner became intrigued by the persistent little boy with the paper route and a desire for the most expensive Meccano kit available. The shop owner offered to special order the set and sell it to Luc at cost plus shipping, for a sum of approximately $360. The order was placed and Luc spent weeks anxiously awaiting his package from England (Figure 6-1). When it arrived it was everything he'd hoped for. He built an eight-foot-tall Eiffel Tower, he built a car that he drove using a blender motor, he built gear boxes. He spent years using the kit, and when he moved to the United States, it came with him. These days his sons use their dad's Meccano Set Number 10 to build homemade Star Wars props and make stop animation movies.

Luc is thoughtful when he reflects on his childhood endeavors:

> *The confidence to take things on also did not come quickly. It took many years to build up courage, there were many discouraging failures—in fact I often felt that every project was a failure, compared to what I wanted it to be. But small successes built into bigger ambitions. This may not be a thrust of your thesis, but it's something that I recognize as important. Somehow I saw the right combination of both, supported by pragmatic tenacity and a growing imagination.*

> *And of course my parents, who never discouraged me, and always supported me. They had a positive and curious outlook on life in general, came from pragmatic countryside upbringing, were demanding of themselves and worked hard in spite of difficulties; they were socially adept, modern thinking, learned new skills throughout life, and exemplified that adaptation was crucial. My dad goes to the Mac store for lessons every week at 84.*

Fittingly, at the end of our video chat, Luc turned his camera to one of the walls in his office at Disney. There, I saw an elaborate prototype for a new ride concept. The model was made entirely of Meccano parts, from the set that Luc bought with his paper route money when he was an 11-year-old boy in Canada.

Figure 6-1. Luc Mayrand, his brother, and his father, with his Meccano Set Number 10 (photo courtesy of Luc Mayrand)

Calling for Help

I was not a tinkerer as a kid. I think I was born more a physicist than an engineer. I marvelled at the way things work. Why things cool off in a predictable way. I think I got to a point that I thought enough about the world, literally everything from thermodynamics to Newtonian dynamics, and realized one day I'm going to have to get a job, nobody's going to pay me to sit around and think and watch the world and try to understand the whys. I'd better figure out how to make the hows so that I can make a living.

And he did. This child who wanted to *"make the hows,"* instead of just *"understanding the whys,"* went on to lead the development of a wheelchair that could climb stairs, a prosthetic arm capable of performing complex tasks, and robotics competitions that involve hundreds of thousands of students annually. So impactful and prolific are his accomplishments that in 2013 this inventor, Dean Kamen (Figure 6-2), was awarded the James C. Morgan Global Humanitarian Award. Not bad for someone who did not consider himself a tinkerer.

Figure 6-2. A sketch of five-year-old Dean Kamen, drawn by his father, Jack Kamen (drawing courtesy of Dean Kamen)

Dean's story isn't one of things coming easily to him. Rather, it is one of a child who was so determined to learn that he wouldn't take no for an answer. He began looking for challenges to work on. It was in the early days of power electronics and the *"whole world was going disco because you could drive a whole building to the beat of music."* Dean became fascinated by the idea of driving lighting systems with sound. He started building sound and lighting equipment that he would then sell to local bands at a substantial markup from his materials' cost. At this point I had to stop Dean for clarification. Had he really taught himself this all by looking at books and trying things? It turned out he did. Plus he had figured out who to call for help.

"I would call the companies and get what we would now call an applications engineer on the phone and I would say 'I'm working on this,' and I'd try to sound like an adult and I would tell them that this was what I needed to do and I was having trouble with their data sheets." Dean admits that it would often take numerous attempts at calling the same company before he would find someone who was interested in helping him. To prepare for the calls, Dean would read data sheets and catalogs. Most importantly he would just try to make things. The more stuff that Dean made, the more people he found who were willing to buy electronics from him, despite his barely being in his teens.

Dean credits a lot of his success to his willingness to take criticism. He would bring his devices to the people who wanted them and, more often than not, they would not be shy about telling him what was wrong with them. After a few iterations, he typically had a device that the client wanted and he would rush back to his basement and start making more of them. Every penny of his earned money went toward buying more electronics, oscilloscopes, power supplies, or machine tools. Before long, the high school student had both mechanical and electronics shops set up in his parents' basement, and he spent long nights and weekends working on his inventions.

Dean admits that he was often frustrated when things didn't work, but he refused to give up:

> As much pressure as I was under to try to get it done and get it done right and make it work, and as much frustration as I had when it didn't work and I had to remake it, I never felt it was as much pressure as sitting in a classroom trying to guess what the teacher wanted as the answer to that multiple choice question. I work on stuff I think is important... If it doesn't work I'll try something else and I'll keep trying something else until it does work. I consider the failures, OK I learned something. It is exciting. I don't know if it's going to work or not. There's no answer in the back of the book. It's new, it's different. If it works it's a big deal. If it doesn't work I'll try something else.
>
> I just didn't like learning by prescription all the stuff that is already known because you can check for the answer at the back of the book. I wanted the answers to the questions that aren't in the book or anywhere else.

Makers like Dean who want to answer unanswered questions, and solve unsolved problems, inevitably run into stumbling blocks. It's how they deal with those stumbling blocks that sets them apart as makers.

Quietly Observing

When Mimi Hui (Figure 6-3) went to her first day of kindergarten, she couldn't communicate with anyone in the school. The six-year-old had spent most of her life living with her maternal grandmother on the island of Macau, and spoke only Chinese. She still vividly recalls her first day of school in the United States as a day of confusion. Mimi did not understand anything that was said around her, and the toys in the classroom made no sense to her. She had never seen building blocks before, and found the nonfunctional toy kitchen sets to be silly. What is the point of a stove that can't cook and just sits there? In short, young Mimi viewed it as *"the most boring day of my life."*

For the next two years, Mimi was silent in school. Unable to speak English, she listened and began to make sense of the words that she was hearing. The ability to understand spoken English came to her much faster than the ability to speak it, so she stayed quiet and soaked in the words around her. Others at her school assumed she didn't understand them, so they spoke freely around her. Unbeknownst to them, Mimi was listening, and learning.

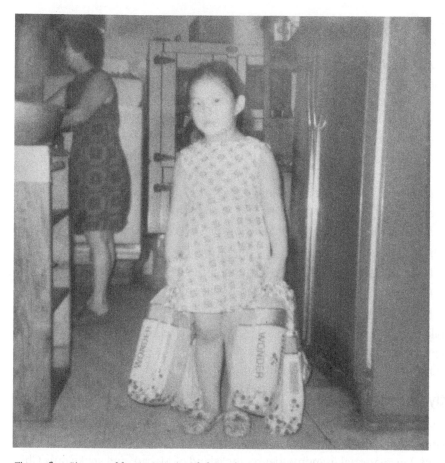

Figure 6-3. Six-year-old Mimi Hui with bags she insisted she could carry alone (photo courtesy of Mr. and Mrs. Hui)

While Mimi was learning the nuances of both American culture and the English language, she spent a lot of time in her parents' take-out restaurant. For as long as she can remember, she was put to work in the restaurant helping with assorted tasks. When she turned nine, her role increased to include taking orders, and within a few years she was the one hiring plumbers and electricians, doing the translations (including legal documents), and helping with the staff.

Despite the work demands on her time and energy, Mimi threw herself into her schoolwork and projects; it was a great escape. When she was nine, her school had a science fair and she decided to research solar cells. She made multiple trips to the library to research solar energy, and made mockups of solar cell designs. (She wistfully notes that the project didn't win an award. Rather, *"the guy with the volcano always wins.*") By high school, Mimi was taking every AP Science class the school offered, and it became apparent that she was talented in this area. Mimi went on to earn degrees in electrical engineering, industrial design, and innovation, and then to engineering roles such as senior programmer at Netscape and management roles such as a senior manager at Frog Design.

One of Mimi's talents is orchestrating the teams of people, and skills, needed to evaluate product initiatives and develop and implement strategic plans. Thus, shortly after finishing her industrial design degree, she founded Canal Mercer Designs, through which she helped both startups and established companies figure out what is needed to get new products on the market. Mimi's career has already involved her working in eight countries and on products ranging from financial systems to reimagining driving experiences for automotive companies. The ability to be thrown into a new team and a new environment, and manage to quickly get to work and bring a product to realization is rare and valuable. When I look at the work that Mimi does, and has done, it's hard not to find myself thinking of a young Mimi, barely a teenager, hiring electricians for a family restaurant in a language and culture that she had been introduced to only a few years earlier.

These days, it's not hard to get Mimi to talk about her views on education. Remembering her childhood confusion about why anyone would think she would find a nonworking plastic miniature stove interesting, and her frustration about there being no making in her school, she has set out to do something about that. She has helped STEM schools in New York design curricula for talented, but at-risk, teenagers and put her product design and engineering skills to work creating kits for kids. Unsurprisingly, there are no pretend parts in her kits, but rather working electronics and experiments involving light and wind theory. *"My feeling is that kids are naturally curious, and sometimes kids don't do well in school because they are bored or they are just not being motivated in the right way, versus they are dumb."*

Not Failing

Persistence is, it seems, a necessary aspect of being a maker. Because most projects involve doing something new, or at least new to the maker, it's not always immediately clear what materials, resources, knowledge, and assistance will be needed. Makers are notorious for taking on hard, or unusual, projects that require multiple approaches and attempts before a working design is developed. This ability to keep trying, even when the first method tried doesn't give the desired effect, is key to success when pioneering new fields and endeavors.

I often hear people extolling the virtues of failure, but I think we should focus more on persistence, and resilience, than on failure. Being persistent doesn't mean failing. It's not failure itself that leads to success; rather, it's the willingness to pick yourself (and your project) back up in hopes of getting it to work. And, if that's what you do, those early hiccups weren't failures, they were rough drafts. Think of the papers you likely wrote in school. If your childhood writing classes were like mine, you were expected to write multiple iterative versions, in which you made corrections. We called them early drafts, not failures. To me, it's only truly a failure if you give up.

I often teach engineering design and graphics to engineering undergraduates (mostly first-year students) and similar classes for teachers. As design becomes something more and more PK–12 schools and colleges are teaching, I'm struggling with the words that I hear people using. I know that at conferences, and among practitioners, phrases like "celebrate failures" and "fail fast" are considered paradigm shifting and eye opening, but I'm really struggling with seeing these words used at the PK–12 level, and even in college classrooms. I have had quite a few educators come and tell me that they want to celebrate failure or that they have heard that teaching children to "celebrate failure" is what they should be doing.

Words matter. When we work with learners who are in difficult situations, it seems disingenuous to tell them to be OK with "failure" and then later in the day if they "fail" the math test we hold them back, call their parents, or take away their scholarship. I liken the design process to writing. We never assume that the first draft is the final draft. We know that we'll iterate and get feedback from users, peers, and others. I have started to say that it's only failure if you completely give up. Otherwise, it's just a draft, or an experiment. Rather than "celebrate failure" or "fail fast," I use "prototype early," "try lots of things and experiment," "learn from your iterations," "try something!" and things to that effect. These words are also more in line with what I see among makers.

One of the great things about Maker Faires is that they aren't competitions. There isn't a loser, or a grade. This frees exhibitors from pressure to have things perfect, or even finished, before showing them. Some of the richest conversations at the faires come when makers find themselves helping each other, and sharing ideas. Just as most makers believe that they can figure out how to make almost anything, it's the rare maker who considers a project truly finished. There's always another version to try, or another maker who has an idea for building off of your project.

Ageless Tools

As I reflect on ways in which makers are persistent—sticking with hard challenges for years if not decades, pushing through challenges and not letting *nos* stop them —I also found myself noticing how many adult makers still use the tools they used as children. Just as Luc still proudly displays his hard-earned Meccano set in his office at Disney, as I interviewed makers many of them had tools and toys from their childhoods still at arms' reach. One such maker is Kipp Bradford (Figure 6-4).

Figure 6-4. Kipp Bradford and a plane he built using LEGO blocks (photo courtesy of Earle and Yvette Bradford)

Kipp grew up in cities throughout the United States, following job opportunities for his father. His parents had been born in New Orleans, but were unhappy with the job opportunities for a black man in the south in the 1960s. His parents stressed the importance of education, hard work, and hands-on skills, and modeled for Kipp the importance of being able to fix and build. Kipp fondly remembers that both of his grandfathers were hardcore tinkerers, and his father was always repairing everything in the house. (That is, until Kipp managed to sneak tools away so that he could work on his own projects.) Despite having access to a wide variety of tools, it was LEGO that captured his imagination. So much so that when I interviewed Kipp in his office at Brown University, where he was teaching a senior-level engineering class, he quickly pulled a LEGO car off of one of his shelves to show me. The car was just one of about 15 LEGO sets within arms' reach of him in his office. Though he still buys LEGO sets, he is also proud to show off pieces that are more than 30 years old, stating *"I still have LEGOs from when I was four years old. I have LEGOs from 1978. They definitely represent a thread through my life of always making stuff. Always being able to imagine things that were interesting to me and then I built them."* Rather than view these as kids' toys that can be mastered or outgrown, they're simply another tool that Kipp uses and becomes better at using.

Projects Without an End

Being persistent is often equated to never giving up. This is a good trait to have, because many makers find themselves working on projects that take much longer, sometimes decades longer, than intended. Many of the people I talked to told me about the projects they've built, and then about the large numbers of projects that are still underway, or yet to be started. Perhaps the best example of the persistent nature of makers is that most of their projects are never truly finished. The award for longest project in progress, among my interviewees, goes to Will Durfee. Although there are many fantastic maker stories that I could tell about Will, there is one that stuck with me.

Dr. Will Durfee has had a distinguished career in engineering education. As a mechanical engineering professor at the University of Minnesota, and before that as a professor of engineering design at MIT, Will has received numerous awards for his inventions and his teaching, publications, and research. But he has a secret hidden away in his closet—a project that he has yet to finish, despite having started it more than 40 years ago.

Will was always interested in space and telescopes. So much so that in kindergarten he impressed the adults around him by building a bright red wooden telescope stand using a handsaw, hand drill, screws, and a hammer. As a child he was given a telescope that *"with a little imagination you could see the rings of Saturn, but not nearly as clear as the photo of Saturn on the telescope box."* He enjoyed bringing the telescope outside and looking through it, and was fascinated by the idea of building telescopes.

A few years later, in middle school, Will read a magazine article about building a telescope. He was blown away by the idea that you could build your own six-inch mirror (which would be a significant improvement over the one in the telescope he owned). He convinced his parents to supplement the money he had saved from his allowance, and ordered the kit.

Just like Chris Anderson and his mail-order submarine kit (or rather, plans) in Chapter 2, Will quickly found that he needed some materials that were unlikely to be found in the family's basement. Key on this list was a 50-gallon oil drum. He managed to convince his parents to both help him find the drum, and let him set it up in his bedroom. Will adds that this placement was essential to the project— had it been stored in the basement he probably would have worked on it less often.

With an oil drum in his bedroom, Will set to work grinding the glass by hand. The middle schooler was quickly impressed by both how well it was turning out, and how slowly it was going. So slowly, alas, that he never managed to finish it. However, Will was so determined to complete this project that he packed up the 2/3 finished mirror in a box and has been carrying it from school to school, and apartment to apartment, ever since. It was with him when he was in Cambridge, Massachusetts, as a grad student, then in his first house when he became a professor at MIT, and it can now be found in his attic in Edina, Minnesota.

The project begun in a childhood bedroom still calls to Professor Durfee, who recently sent me an update:

> There are two pieces of good news. First, glass does not deteriorate, so I can immediately pick up where I left off. Second, now that I have added embedded microcontroller skills to my toolbox, I can build a super-cool equatorial mount stand that auto-tracks celestial bodies. Yes, I know you can buy these, but much more fun to build.
>
> Building a telescope stand will almost take me full circle as apparently, in kindergarten I wowed the teacher by building, out of wood and painted bright red, a fixed-angle, wheeled telescope stand. Maybe I should just go with my strength and build the stand first.

One of the reasons this story sticks with me is because, even now, it is evident how much this project means to Will. The first time I heard about it, we were in a coffee shop on the University of Minnesota campus, and the smile on his face while remembering his young victory of convincing his mom to let him keep the oil drum in his bedroom was telling. This project clearly captured part of his imagination then, and his painstaking packing and moving of the mirror box from home to home and state to state strikes me as an example of a slow, appreciative, persistence. He has gone from a kindergartner building and painting a bright red telescope box, to a full professor who still spends time contemplating how to use the skills he's nurtured in the years since then to improve the original project plans. Despite accolades for his work in stroke rehabilitation and hydraulic actuation systems, he has never forgotten, or truly abandoned, his childhood telescope lens.

The Many Faces of Persistence

Dictionary definitions of "persistence" focus on the quality of continuing on in the face of opposition, discouragement, and obstacles. These obstacles look different for every maker, but the ability to overcome challenges is something that every maker I've met has in common. Whether it was a lack of materials, a lack of funds, challenges understanding the language or culture they were dropped into, or ignoring people who told them what they were trying to do wasn't possible, these children found ways to accomplish what they set out to do. Not every young maker is as confident as Luc Mayrand, who believed that as a child *"[he] could push [him]self to do anything,"* but each of these makers demonstrated their ability to continue, if not to thrive, in the face of adversity.

Strikingly, none of these stories involved parents, teachers, or other adults jumping in and removing the obstacles. Adults and mentors were around and supportive of these makers, but rather than simply buy the desired tool to do the hard parts of the project, they encouraged the child to keep at it and find ways to solve their own problems. Sometimes it was the mother who let a middle school student keep a 50-gallon oil drum in his room; other times it was a hobby shop owner who offered to sell a kit at cost. Each of these actions showed the young makers that they were on the right track, and set them up for the next step, but made them do the heavy lifting themselves.

As a parent or mentor, it's sometimes tempting to remove all obstacles from a child's path. Doing that, though, removes the chance for our children to see how strong, and how persistent, they can be. The makers that I talk to are still visibly proud of the things that they accomplished as children, even more so when they recognized that they were doing something hard that would require many tries before they'd get it right.

7

Resourcefulness

Makers are resourceful.

They look for materials and inspiration in unlikely places.

While preparing a presentation for a group of young adults and their mentors, I found myself looking through pictures of one of my mentors, Paul McGill. Paul, an electrical engineer at the Monterey Bay Aquarium Research Institute (MBARI), designs and builds cutting-edge research tools for underwater exploration and monitoring. Given that my time as an intern at MBARI was spent in the robotics lab and asking Paul for lots of advice, I was quite happy to find a picture of him on a research vessel in Antarctica (Figure 7-1), standing with other engineers and the remotely operated vehicle (ROV) that they had designed and built for the expedition. I thought that this would be a great image to end my talk, and gave me a chance to highlight the incredible engineering that he does. I emailed Paul asking for per-mission to use the image, and he quickly replied with a yes. He then asked me if I knew the story behind that picture. I simply thought that it was a beautiful picture of Antarctica, engineers that I knew and respected, and a nifty robot. It turns out I didn't know just how nifty the robot was.

The robot in the picture is not the one that the MBARI team had worked on for months prior to the expedition. During one of the first dives from the icebreaker, the original robot was sucked through the giant propellers of the ship and never came up again. Paul recalls being on the boat, with a full contingent of scientists, 40 more days in the Antarctic, and no robot. This is the point where many people would give up and go home (or take a lot of pictures of icebergs). Not Paul. He and the other two engineers took an inventory of their few remaining parts. They had a spare underwater camera, a couple of electric thrusters, and the control electronics that stayed on the ship. But they were lacking many critical components, including the ROV frame, underwater electrical connectors, and pressure-tolerant flotation to counter the ROV's weight. The engineers started scouring the ship, scavenging for parts, and creating the missing components from pieces of scrap. Three days later they had an operational ROV, ready for diving. This hodgepodge of a robot was able to dive into the frigid Antarctic water for weeks and was able to collect nearly all the data originally hoped for from the expedition. High-performance, cold-water ROVs are hard to build in a research lab. This is the first one that I've ever heard of built entirely on a research vessel out of scrap parts.

Figure 7-1. MBARI engineers working on "Phoenix," an improvised ROV for studying free-drifting icebergs near Antarctica (photo credit © 2008 MBARI)

Perhaps the most surprising part of this story is that I wasn't really surprised. One of the reasons I so admire Paul is that he is the sort of maker who can seemingly create anything out of found objects and a soldering iron, smiling during the entire process. As you can probably imagine, this isn't a personality trait that suddenly appeared in adulthood. Paul has a long history of recognizing the treasure in other people's scraps. As a child, Paul used to walk down streets looking for repairmen on telephone poles. When he found one he would ask them nicely to throw down some copper wire so that he could use it for something he was building. Amazingly, many of the workmen did just that and Paul happily incorporated the wire into his projects, including a telegraph that he built in fourth grade by winding the wire around iron nails. Whenever he found something that might be interesting, even if he had no use for it at the time, he would stash it away, thus building up a collection of spare parts and widgets.

Being able to look at things and imagine ways that they can be used is a skill that Paul was taught by his mother. As a single mother, she often didn't have the money to call a repairman, so instead would take Paul (Figure 7-2) to the hardware store. Her rule, though, was that they weren't allowed to ask for help. Instead, they had to walk down every aisle and look at all of the parts and try to figure out which they needed. When they saw something unfamiliar, they would ask themselves "what could we use this for?" They always left the store with more things than they went for. Paul's mother nurtured his curiosity by insisting that he look, inspect, and figure out things for himself. Through their "what could we use this for?" game at the hardware store, Paul learned to see multiple possibilities in every object.

Figure 7-2. Eleven-year-old Paul McGill (photo courtesy of Paul McGill)

When Paul visited his father, an engineer, they would also work on projects together. When Paul got his first soldering iron for his 13th birthday, his father taught him how to solder components together to build simple circuits. Most of the circuits didn't work the first time, and Paul's father would say, "That's OK, let's figure out what went wrong." They would carefully inspect the work to find the component that had been overheated or the tiny bridge of solder that was creating a short circuit. Paul learned that patience and the willingness to keep trying are an important part of the creative process.

Paul was an avid reader, once convincing an elementary school teacher that instead of a book report he would build a working diorama of an Antarctic research station complete with tiny electric lights (Figure 7-3). Thus it's not surprising to me that the little boy who walked down the street scrounging copper wire, taking scraps from construction sites, and imagining ways to build things with the parts at hand would grow up to be part of a team that, almost magically, turned spare parts and scavenged materials into a functional underwater robot, saving the day for an Antarctic science expedition.

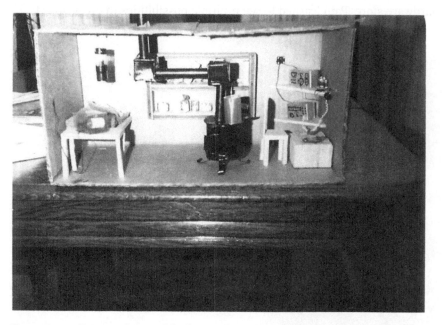

Figure 7-3. A diorama that Paul built, complete with flickering lights, based on the book Alone: The Classic Polar Adventure by Richard E. Byrd (photo courtesy of Paul McGill)

Purposeful Repurposing

"Our family didn't have a lot of money when I was growing up, so we had to make our own playthings." If there is one phrase that I heard most often over the course of doing interviews for this book, it was this one and its variants. As put by Luc Mayrand, who we met in Chapter 6, *"Lack of money is oftentimes the source of invention."* When you have the funds to buy whatever you need, or think you need, it is all too

easy into the trap of believing that if you don't have the "right" part, you simply can't complete a given project. When you know there is absolutely no way you can afford that part, though, and you still *want* to do the project, things get interesting. Many makers derive great joy from combining things in unusual ways and finding unexpected uses for common tools and objects. Makers who grew up responsible for entertaining themselves as children and finding and creating their own toys approach problems differently as adults. They are more likely, it seems, to act like Paul McGill and his mother in the hardware store: take stock of what is around you and figure out how you can make it do what you need.

In almost every class that I teach, I meet students who claim they can't complete a project because they don't have the correct parts. I have to wonder if maybe we've made it too easy for them. If you grow up always assuming that a "correct" part exists, and that things should snap together easily, you're not used to thinking outside the box and imagining ways to make the "incorrect" parts do interesting things. The stories in this book often stem from situations where unusual combinations and ingenuity lead to entirely new ways of approaching problems.

For some makers, such resourcefulness starts at a young age. One day when Raquel Vélez (Figure 7-4) was four years old, her mother noticed her holding a large bunch of screws in her hands. When asked, Raquel told her they were from school. Sensing that there had to be more to this story, her mother insisted that Raquel return them the next day. When Raquel returned the screws her preschool teacher was rather relieved. She had been wondering why the classroom's air conditioning unit was missing all of its screws. It turns out that Raquel hadn't felt like playing with her classmates the previous day and had quietly snuck off to see if she could remove the fasteners. When I asked Raquel whether her classroom had made screwdrivers available to the students, she laughed and replied that she must have managed to find a piece of metal to use as a makeshift screwdriver or else had resorted to using her fingernails.

Raquel's ability to quickly find, or develop, tools to accomplish her goals has only grown since preschool. As an undergraduate at Caltech, where she studied mechanical engineering, she became active with numerous groups working on innovative robotics applications. She founded and led Caltech's student robotics rescue team that developed an urban search and rescue vehicle. After graduating, she proved herself quite gifted at learning, and applying, new skills by working for companies and organizations such as MIT's Lincoln Laboratory, Applied Minds, and Storify. (In one particularly striking example, Raquel managed to go from brand-new intern to junior developer in five weeks at Skookum Digital Works, where she developed custom web applications.)

Figure 7-4. Two-year-old Raquel Vélez explores a drawer (photo courtesy of Germán Vélez and Deborah Aguiar-Vélez)

These days, she can be found combining her loves of robots and software as one of the leading players in the NodeBots. NodeBots lies at the intersection of JavaScript and Robotics, and is a way for programmers to easily start writing code that interacts with the physical world. Says Raquel, *"I think my favorite thing about NodeBots is that I get to use my hands, break things on purpose, and put it all back together (hopefully better than when I found it!)."*

Books as Bricks

As I mentioned earlier, many young makers were avid readers. Bradley Gawthrop (Figure 7-5), a pipe organ designer, was no different. The Gawthrop children grew up in a household full of books. *"When our parents got married, two of the world's greatest libraries merged."* Books are great for gaining knowledge but also, as Bradley noted, are great building materials. Towers of books were turned into furniture and forts. Encyclopedias were particularly useful as giant bricks. Not content to stick to books as a building material, Bradley and his four siblings would scour the neighborhood and the house for supplies. *"We were hoarders of the first order."* Bradley explains that his and his siblings' resourcefulness was driven by necessity. They were a family of seven living in a three bedroom house and money was tight. The family lived adjacent to the Quantico Marine Base in Virginia, so many of their playmates were other military kids. Many of these families were very hands-on and could be found fixing things around their houses. Growing up in this community, Bradley took it for granted that people knew how to build things and do things. It wasn't until he got his first job that he realized how unusual some of his childhood experiences were.

Having five children in the family also meant that large-scale projects were possible, sometimes beyond what their parents were expecting. When they innocently asked their mother if they could dig a hole in the backyard, she agreed. However, she hadn't accounted for five kids working in shifts in the backyard. *"It's amazing what you can do if you're bored enough."* By the time their mother checked on the project, they had dug trenches the size of military foxholes. The kids even worked out a lookout system that they would put in place for projects that weren't parent approved, or which they, usually correctly, assumed would get them in trouble. With some regret, Bradley admits that the backyard construction site was promptly filled in at their mother's request. Amazingly, she continued to allow her children fairly unrestricted access to tools, which *"led to [their] not being afraid of tools and not having the mentality if something isn't available the way you want that it isn't available at all."* Though Bradley did point out that *"the problem with giving kids real tools is that they are apt to use them."*

The Gawthrop children were homeschooled, so Bradley was able to start working at 16. As an 11-year-old, he read a book on pipe organs and was fascinated ever since. Bradley is a firm believer in learning by doing, and from that point on worked in a number of jobs where he gained skills that he deemed as possibly useful for his dream job. These days, Bradley has achieved his childhood dream and runs Gawthrop Organworks, a pipe organ building firm, restoring old organs and building new ones from scratch. When collaborating with and training others, Bradley has been surprised by the lack of mechanical understanding many people, even those with engineering degrees, have about cause and effect and how things work. This would often result in their creating things that looked beautiful in a model on the computer, but could not be built. Even without formal training in engineering,

Bradley had a life of experiences with tools, apprenticeships, and simply taking things apart and figuring out how they could be put back together.

Figure 7-5. Bradley Gawthrop works on electrical cabling for an organ he built in Boone, NC (photo courtesy of Bradley Gawthrop)

Tooling Up, Powering Down

Holly Gates works on things that clearly fall in the "very high tech" category. He's spent much of the past five years developing techniques for enhancing the surface texture of solar cells in ways that will enhance their light-trapping capabilities. Holly and his team also design and build the process, machinery, and materials needed to bring these techniques to reality. He has a long history of designing cutting-edge technology. As an undergraduate at MIT, he worked in the research lab that created E Ink, the technology that you now find in your eReaders (such as the Kindle) and many retail displays and wearables. When E Ink spun out of the MIT Media Lab, Holly followed, helping design hundreds of printed circuit boards that would help make these devices possible.

At E Ink, Holly became known as someone who could bring ideas to life and spent a lot of his time creating demonstrations of their technology. As anyone involved in early-stage technology knows, getting investors and early adopters depends on their ability to see your technology in action and being able to imagine the incredible things that it would make possible. Alas, another aspect of early ventures is that you need to create these mindblowing demonstrations at as low a cost as possible. Thus, demo creation is an art that requires technological mastery coupled with an eye for frugality. This became one of Holly's signatures. For a young, creative engineer, working to make an idea a reality is a dream position. *"E Ink was a great opportunity for that because the basic technology was so different and cool that even with relatively simple demo efforts you could make something really impressive. Of course my role would have been impossible without all the hard work and true innovation required to make the display material itself, but it was immensely satisfying to be in a position to bring the ink the last step of the way from a gray plastic sheet to a vivid paper-like electronic display."* Keep in mind that at the time Holly was working on this he was so early in his career that he left his master's program to work at E Ink. Yet, his impact can now be seen in many of the devices we now take for granted. As with all of the makers I've met, for me the burning question is *"How did you learn to do this sort of stuff?"*

Over the past three years, I've heard some amazing stories. However, I think Holly's childhood home wins for most unusual. Holly's father had grown up wanting to be an airplane mechanic, but his mother wouldn't pay for the training. So he decided that he'd learn to build boats, and went to a boatyard where he could find a job as a laborer. Over time he worked his way up to foreman. During this time he purchased a boat hull and leased a spot in the boatyard where he could store and work on it. When Holly was born in 1975, his parents were living illegally in a tarpaper shack in a boatyard in Costa Mesa, California. Attached to the tarpaper shack was a 32-foot boat that they were building. To get power, the couple ran an extension cord and water hose from a nearby building. When their newborn needed a bath (Figure 7-6), they filled a galvanized tub that they had put on wooden pallets outside the shack. Holly's earliest memories are of playing hide-and-seek in *"enormous stack of old tires in the boatyard."* Holly's parents had planned to sail their small family around the world, but his parents separated before that could happen. Holly and his dad moved into the boat, moored in the harbor, until he was five.

Figure 7-6. A young Holly taking a bath in the boat's living quarters (photo courtesy of Holly Gates)

Holly's childhood and later life is one in which resourcefulness plays a starring role, both in his day-to-day work life and in how he and his wife are raising their three children. For his middle school science fair project, Holly used instructions from a book that he got in a used bookstore he often frequented with his mother. He then scoured his town looking for an old neon sign. He found an unused one at a stationery store and set to work dismantling its 9kV transformer. After recovering from a *"solid zapping"* from that transformer, he set to work constructing his project. *"[The zapping] was an experience I won't forget, but fortunately the current limit on that transformer (as on most neon sign transformers) was below what is generally considered deadly."* After much work, and more scouring for parts, Holly finished a homemade Tesla coil (Figure 7-7), which could make arcs half a meter long that could be drawn to metal objects that spectators held in their hands. Unfortunately, bringing this to the county science fair proved disappointing when the judges refused to let him turn it on, citing safety concerns. Even more annoying is that Holly still isn't completely sure whether he managed to convince the judges that it really was his project, not his dad's.

Figure 7-7. Holly and his homemade Tesla coil (photo courtesy of Holly Gates)

While Holly's and my time as undergraduates at MIT overlapped, it wasn't until last year that I knew who he was. As I became more and more interested in making my own clothes, and dabbling in pattern creation, I started reading blogs of other parents who sewed for their children. One of the blogs that I stumbled across was "Tooling Up." This blog, written by Holly, is a detailed account of projects that he, his wife, and their children have taken on. Don't go to this blog looking for microcontroller projects or the latest on 3D printing. Rather, Holly documents the design and creation of Regency dresses, home-cured bacon, hand-ground flour, and handmade straight razors for shaving. I suspect that I'm not the only reader who is surprised, at first, to find that the father who hasn't bought new clothes in three years is a photovoltaics engineer with an electrical engineering degree from MIT and a decidedly high-tech day job. While he's mending clothes with a treadle-powered machine, it's possible you're reading this book on an eReader that uses technology he helped develop. Honestly, I found this surprising, and asked him about why he is drawn to using such old technology when he has access to more cutting-edge devices. His answer gave me a lot to think about:

> [I] like the idea of reusing old stuff when it is feasible to do so, rather than buying new stuff. And I like to learn about, and if possible, experience how the things that surround us are made. The ways in which things are made now are usually not possible to replicate at home or on a small scale. But if you go back a few generations in the production methods, it is sometimes feasible to try it yourself. To tell you the truth to a certain degree I'm not entirely sure why I'm attracted to these antique and low-tech methods. For some reason I find them cool and rewarding. Possibly I'm proving that as people get older, they inevitably get weirder and more obsessed with the past.

To clarify his point about some of the merits of using older technology, he describes how he trims the bushes in his yard:

> I used to cut my front hedge with an electric hedge trimmer. This meant I had on earmuffs, had to run an extension cord and wind it up afterwards. It was loud, produced a ton of vibration, and was annoying and perhaps a little dangerous for all the foot traffic on the sidewalk. Now I use hand shears, which do take a little longer and don't make as precise of a hedge. But it's amazing how many people stop to exchange a few words with me while I'm trimming, and my kids can come hang out and try to clip a few leaves themselves.

When I originally began thinking about resourcefulness as a feature of the maker mindset, I'll admit that I was focused solely on technological resourcefulness. However, as is evident in many of these stories, one of the greatest resources that makers have is each other. It's a nontrivial skill to develop ways of meeting new people and learning their abilities, and as I reflected on it, it became apparent to me that this hedge trimming example of Holly's highlights the human resource side of the Maker Movement. What has made this movement so successful is the willingness for makers to connect to each other, online and in person. Part of this is knowing when it's time to unplug, literally and figuratively, and open yourself up to new experiences and people, because you never know what new connections and skills will prove useful in the future. Who knows, it's entirely possible that one of the people who stops to chat with Holly as he's trimming hedges may become a co-conspirator for some amazing future project, or a lifelong friend.

High Tech, Low Tech, No Tech?

Holly's comments and stories also helped me pinpoint one of the tensions I see between what my first-hand experience with the Maker Movement has been, and how it is portrayed in most of the media. To me, the Maker Movement is about the makers. It's tool agnostic. It's about people who make things, and who love sharing what they make. However, one of the concerns that I often hear from other educators and parents is that "we can't do that stuff, it's too expensive." Further explanation typically reveals that they are equating making with the necessity to buy a 3D printer and a laser cutter and... insert whatever the latest new technology is. While most makers that I know love learning about and trying new technology, what drives them is the love of making things, not a particular piece of equipment. Some makers' tool of choice might be a 3D printer, but others prefer hand tools. Some are using digital fabrication, and others are figuring out jaw-droppingly ornate ways to use cardboard. A child who whittles a wooden sculpture is just as much a maker as one who creates a remote-controlled humanoid robot.

One maker who views the "high" and "low" tech worlds of technology, craft, and art as intricately intertwined is Dr. Leah Buechley (Figure 7-8). As the founder of MIT's "High-Low Tech" research group, Leah is a pioneer in areas such as e-textiles (sewing with electronics), and paper electronics. A visit to her studio is like entering an artistic version of Willy Wonka's Chocolate Factory—patterns on walls change as you pass by, giant cardboard "books" open up into pop-up architecture that you can walk through, pens paint functioning electrical circuits, pop-up paper sculptures light up and move, and all around are examples of fabrics with embedded electronics. One of the things that I find so inspiring about Leah is that she brings her design aesthetic and interest in new technologies to a dizzying array of older art forms such as textiles, paper cutting, and bookmaking. She is the inventor of the Lilypad Arduino, which is the first open source sewable microcontroller board. The Lilypad revolutionized etextiles, particularly in the context of education. Schools, museums, and artisans around the world now use this tool.

Figure 7-8. Leah Buechley in her studio (photo courtesy of Larry and Nancy Buechley)

The child of two furniture builders from northern New Mexico, the world around Leah was one that her family was constantly creating for themselves. Her parents built their studio, and did all of the plumbing, wiring, and finish work in their house. Leah grew up *"saturated in this ongoing making of everything from food and utilitarian stuff, to fixing things. We needed a plow for the road to get to our house in the wintertime, so my dad built a plow."* (Figure 7-9) While money was tight in their family, her parents never let the need to make and fix things seem like a hardship. *"I admired my parents for being able to do that well. They did a good job of communicating to me and my brothers that this was an enjoyable and rewarding thing for them."* Their circumstances meant that the entire family had to be creative and imaginative with the materials that they had around, and young Leah saw this as a fun challenge. Tool use was an expected skill in her family, and she would work and play in her parents' shop. As they worked on full-size furniture, young Leah would design and build doll furniture, which she sold alongside them at craft shows. While nails, hammers, and screws were introduced to her at about three years old, her parents made being allowed to use tools like the band saw a family rite of passage as she got older.

Figure 7-9. A young Leah gardening (photo courtesy of Larry and Nancy Buechley)

Despite being interested in crafting, Leah opted out of taking classes such as art, home economics, and shop in high school, opting instead to study music. (She did note that one of her female friends chose to take shop class, and that it was considered scandalous at the school for a girl to enroll.) Throughout Leah's entire school career, her interests would wander back and forth between mathematical fields and the arts. So much so, that while she ended up majoring in Physics, she started college studying dance and theater and considering a career as a dancer. Over time she found she was drawn to computer science as she saw it as a discipline that let you be *"be creative with mathematics."*

Graduate positions in Leah's lab at MIT quickly became sought after by designers and engineers interested in rethinking the materials and methods that we use to do computation. Given the very rich background that Leah has in traditional crafting and arts, I was curious about whether the students she brings into her group already have that experience. While she says that the experience level varies among her students, she looks for those who have a background in creating something, regardless of what it is. While young adults who have little to no experience with complex manual tasks often struggle picking up those skills, those who come to her with a background in any making domain have a relatively easy time learning new techniques. They also tend to be the ones who can see the possibilities in the materials in front of them.

Learning Resourcefulness

I once found myself in a conversation with a group of fantastic educators who had developed a curriculum that used robotics experiments to explain concepts in a calculus class. They found that their students were often frustrated when their experiments didn't match the theoretical predictions exactly. The solution that the instructors were pursuing was to find higher-quality components, such as more precise gears, so that the experimental results would more closely match the theory. Given that the students in the class were engineering students, it makes me wonder if we sometimes try too hard to give children and students experiences that are too clean and predictable. One of the realities of robotics, and virtually all physical creation, is that there are often initially unexpected results. The challenge for designers and makers is to find reliable ways to deal with the messiness inherent in building real systems. This requires an adaptability and resourcefulness that doesn't come from only using that which have been preselected for their accuracy. More often, the insight that allows a project to function the way the designers hoped comes from reaming up an entirely new way of using tools and knowledge. And that insight is likely to come from someone who has a history of thinking outside the box. Or sometimes it comes from literally using the box... as a building material, as a stepstool, or as a roof for a fort made of a childhood set of encyclopedias.

8

Generosity

Makers share—their knowledge, their tools, and their support.

As the mother of two kids under the age of six, I feel as though half the conversations I have with them are about sharing, cooperation, and collaboration. Why we share, why we take turns, why you should let your sister play with your paints, and so on. I suspect that I'm not the only parent who feels like this is a lesson their children hear constantly. Or, more likely, it feels like something we say quite a bit but that our children seem not to always hear. I often wonder, though, whether we adults practice what we preach. Do we share our time? Do we offer advice, or more importantly, answer the questions that are asked of us? Generosity can take on many forms. The Maker Movement celebrates the sharing of knowledge, as well as the sharing of tools and time. Many makers learn their skills through studying, formally or informally, with other makers. Very few projects are truly solo efforts, and makerspaces are becoming hubs where people can help each other on projects big and small. Makers, new and seasoned, are encouraged to try new things and ask lots of questions. There is rarely shame in not knowing the answer. The important thing is knowing who to ask for help. There are no grades, and makers are encouraged to give away the "secrets" of how their projects work so that others can build off of them.

Contrast this with some of the standard ways in which kids are taught. We spend a lot of students' formal education telling them to "keep their eyes on their own paper," and grading them every step of the way. Within two months of starting kindergarten, my daughter had taken her first standardized test, complete with practice tests and privacy folders so that students couldn't copy each other's work. This is the exact opposite of how most real-world projects are undertaken. Outside of a classroom, almost every project involves working with diverse teams and getting feedback from others along the way. The final product is what matters. When a bridge collapses or a car malfunctions, this end result is what matters, not whether the first prototype worked. We want to find the bugs and challenges as early as possible, and to do that we want as many eyes as possible on the work. Even better, sharing those bugs and challenges with others provides a chance to teach, and to hopefully save other makers some frustration. Within the Maker Movement, information sharing is encouraged and "black boxes" discouraged. It's a rare Maker Faire where you don't see at least one person wearing a shirt that says "If you can't open it, you don't own it."

Open Source

Nathan Seidle (Figure 8-1) grew up making things and taking things apart. As he put it, *"I always learned better with my hands."* His parents encouraged these endeavors *"as long as there were no scars, no physical evidence."* From a young age he spent time in the garage hammering and doing small projects on a work bench that his father had built for him. Despite Nathan's habit of often misplacing the family's tools, Nathan's father encouraged him to use tools and take things apart. When he was about nine, the Weed Eater broke and Nathan and his father dismantled it. They removed the motor and spindle, propped it up on bricks, and plugged it in. (His dad was standing next to the circuit breaker, just in case.) Together they watched it speed up and run across the garage.

In high school, he had a fancy graphing calculator. Given that he was a talented mathematics student with a strong interest in computers, it's not surprising that he wanted to find a way to connect his calculator to his computer. Such cables were commercially available, but they weren't cheap. So Nathan went on a BBS (bulletin board system), a type of online message system that was popular in the 1980s and '90s), and started looking for information on how to make such a cable. He found a schematic online that he used to build a functioning cable (which is particularly impressive when you consider that the BBS was text-based and so was the schematic). This was the first project of this type that Nathan had undertaken, so he didn't even know where to buy the parts. Other posters on the BBS advised him to buy them at RadioShack, which he did, thus beginning a small business of building and selling these cables to his friends at a price much cheaper than the commercial rate. (When I asked if the cables worked, Nathan's reply was an enthusiastic *"A few of them did."*) While making his first cable he accidentally got solder on two of the connector's pins. He used a box knife to remedy this and ended up cutting himself, leaving a scar. Now he looks back and thinks *"that's the scar that started SparkFun."*

Figure 8-1. A young Nathan Seidle prepares to take apart a hotel door to figure out how the lock works (photo courtesy of Dana Seidle)

As I heard this story, I was immediately struck that Nathan's first entrepreneurial endeavor involved creating something from plans that were freely available to anyone willing to take the time to follow them. What sounds at first like a somewhat crazy business plan, proved incredibly successful. Today, Nathan Seidle is the founder and owner of SparkFun Electronics (*https://www.sparkfun.com/*), a hardware company that has more than 155 employees, and a revenue of $30 million in 2013.

SparkFun has been one of the companies at the forefront of the open source hardware movement, with Nathan helping write the original definition and serving on the Open Source Hardware Association (OSHWA) board to promote its benefits. As defined by OSHWA, *"Open source hardware is hardware whose design is made publicly available so that anyone can study, modify, distribute, make, and sell the design or hardware based on that design."* As counterintuitive as it might seem at first, SparkFun is a company that has become wildly successful by making things that have no patent and which, in theory, anyone could legally copy.

When I asked Nathan why he started a company focused on open source hardware, he replied that *"Open source stems from my feeling of not having confidence in myself. When I released my first products in 2003, it was stuff I designed and I wasn't sure if it would work for the other customers. To head off some potential technical support issues I supplied the [schematics and source]."* As SparkFun grew, Nathan began to see other benefits of the open source nature of its products. *"Now we use open source as a way to keep pace by making all of our stuff open. Anybody can copy us so we have maybe 8 to 10 weeks to sell a product before we come up with something better. If you have patents you can sit on your laurels... If you're open source you've got to be sharp."*

SparkFun invested heavily in building up its education team, with 8 of its 155 staff members housed in its Department of Education. Nathan is particularly passionate about the educational benefit of open source hardware. *"Open source in education allows people who are willing to look more opportunities for content... I think open source leads to a lot more remixing and innovation because people can learn off of others."* Nathan is careful to point out that he is not an educator, but SparkFun has built a strong reputation for working with formal and informal educators. Its commitment to sharing is made even clearer by a note at the bottom of its education website: *"Sharing Means Caring. Everything you find at SparkFun is free and usable by anyone. All that we ask is that if you see a place for improvement that you let us know."*

For many children, show-and-tell is a highlight of the early school years. It's a chance to show off the neat things they found or made. It seems fitting, then, that the tagline for Maker Faire is "The Greatest Show-and-Tell on Earth." Having been to many Maker Faires, both large and mini, I can attest to the truthfulness of that tagline. Seeing makers showing a sense of pride and eagerness to share their work is quite reminiscent of the excitement that grade school show-and-tell days had. True, there weren't many lasers or robots at my elementary school's show-and-tells, but the idea of having a space for excitedly sharing things that you find important is definitely present in both events.

The idea of freely sharing your projects, as well as detailed information on how to replicate them, is a large part of the Maker Movement. Whereas makers a hundred or more years ago would have had to meet in person, in settings such as sewing circles or clubs, or share their work by letter, the Internet allows makers around the world to connect and share ideas almost instantly. Makers often post their projects, complete with step-by-step directions and usually a discussion of the challenges they faced, on sites such as Instructables (*http://instructables.com*) and MakeProjects (*http://makeprojects.com*). Another site, DIY.org, was created as *"a place for kids to share what they do, meet others who love the same skills, and be awesome."* Skills include backend developer, baker, cardboarder, circuit bender, fashion designer, fort builder, and wind engineer. DIY is set up specifically for young makers, creating an environment where their identities aren't revealed and projects and skills are geared at kids. In many ways, sites such as these are an updated version of the BBSs that a young Nathan Seidle went to for project plans, or computer programming magazines, complete with code samples to be retyped, that other young makers in the early 1990s persuaded their parents to order for them.

Sharing Bites, Not Bytes

While a young Nathan Seidle got his start in entrepreneurship selling calculator cables, Nick Kokonas got his by forming "The People Who Bring You Games" with his friends at the age of 12. This was the year that his family got an Apple II computer, which he would stay up all night programming, eventually writing accounting software for his dad. Even now, decades later, it doesn't take him long to find a container of old floppy disks, modified such that he could use both sides of them, which he shows me proudly. A child's handwriting proudly labels the various games with titles such as "Beer Run from the People Who Bring You Games!!!" These weren't, however, games that he and his friends had written. Rather, after realizing that he could bypass the security measures on early games, Nick and his friends would buy games and then make copies that they could sell at a fraction of the cost. To this group of kids it was a fun challenge. Upgrades in encryption were seen as puzzles that they needed to solve, and it allowed their classmates to play games they wouldn't have been able to afford. Nick would grow up to use his creativity, technical skills, and entrepreneurial spirit in a completely different field than video games.

I first learned of Nick and his work while reading the book *Life on the Line*, (Gotham) which he wrote with chef Grant Achatz, his partner in creating Alinea restaurant in Chicago. I had picked up the book at the library for fun. However, as I read this story of two men, one who as a child cooked in his family's restaurants and rebuilt a car with his father, and the other who, as I later learned while interviewing him, spent long teenage nights on a computer writing code, it became clear that they were classic makers. Thus, it's unsurprising that the restaurants and bars they've created together are unique.

The entrance to Alinea is unlike any passageway that I have ever encountered. A dark perspective-shifting hallway floored with dirt and grass greets those who enter the restaurant's unmarked doorway. At the end of the narrowing hallway is a wall of kitchen skewers that would lead to an uncomfortable start of the meal, were it not for a side doorway that visitors can duck into before reaching the spiky dead end. Alinea, and later Next: restaurant and Aviary bar, was a striking departure from most restaurant traditions. New techniques were used to make the food, new tools (such as an antigriddle that flash-freezes foods on a –30 degree Fahrenheit surface) were developed, and flavors were played with. They even developed their own ticketing software through which patrons would make reservations in a manner more akin to buying a theater ticket than a traditional restaurant reservation. There is a strong sense of whimsy, and questioning the usual ways of doing things, in almost every aspect of Nick and Grant's endeavors.

In the movie *Spinning Plates*, Grant explains that *"Alinea is a place where food is at once art, at once craft, and at once science."* That combination—art, craft, and science—could describe the work of most makers. The idea of combining artistic expression, skill, and technology is relevant whether we're talking about synchronized musical aerial robots or edible floating balloons. Over the course of my discussion with Nick, other parallels between the work that he and Grant do in the realm of hospitality and the open source hardware community became apparent. While the idea of open source hardware is still fairly new, the idea of publishing the "recipe" for something you sell is not. It's not coincidental that books of prewritten software code are often called "cookbooks" or "recipe books." Many world-renowned restaurants and chefs produce cookbooks that allow the users to re-create, or at least try to re-create, their famous dishes at home.

When Grant and Nick set out to create a cookbook of Alinea's recipes, they were adamant that it be a book that allowed for faithful reproductions of the restaurant's food. They decided that if they were going to publish a cookbook, they would publish the recipes as they use them in the restaurant, and include pictures of every dish. They wanted people to be able to make the recipes at home, or even make specific elements of a plated dish:

We didn't use a recipe tester, because what does a recipe tester do? They tell you what can't be done at home. ... You get a recipe cookbook and what you don't realize is that none of the recipes are the actual recipes, because no [restaurant] in America does cups and teaspoons and tablespoons. They do grams. And now you don't need to worry about the density of it, because if it's 40 grams of water, you put it on a scale and it's 40 grams of water. If I need 8 grams of bee pollen, it's not "is it ground;" it doesn't matter. It's 8 grams of bee pollen. We basically said we're doing it all in metric, you're going to have to buy a scale, and there are a few things you need if you want this to come out the way it comes out in the restaurant. But I don't need to hire a recipe tester because I know [the recipes] are right since they're the same ones we're making in the restaurant every night.

In the case of Alinea, they took open source even one step further. They published their business plan online, from layout to logos, while they were building the restaurant. Nick said that this attitude toward openness can be traced back to something his father told him when he was young: *"Don't worry about hiding anything. If you get the best ideas in the world, you could go on a hilltop and shout them out, and 99% of the people will tell you you're an idiot, the other 1% will say 'that's brilliant, I'm going to copy it' and they won't bother doing it because you've still got to show up every day."*

While Nick's father stressed openness to him, his maternal grandfather got him using tools. After coming to the United States alone as an orphan at the age of 14, Joseph Szwedo found a job preparing pathology slides at Children's Memorial Hospital, where he worked for more than 50 years. As a hobby, though, he taught himself to make violins, despite not being able to play the instrument at all. He became so skilled at this that he was sought out by musicians from the symphony to do repairs, which he did free of charge. It was about making the right pieces, not getting paid. He had a basement shop where he would let Nick play among the smells of tobacco smoke and varnish. *"Here's a pile of nails and a board and a hammer."* Nick went on to do woodworking throughout high school and college.

Before I left our interview, Nick introduced me to his sons. One son, who is approximately the age Nick was when he formed "The People Who Bring You Games," creates elaborate live-action re-creations of video games complete with costumes, scripts, and special effects. Just like his great-grandfather, James taught himself the skills he needed to pursue his hobby. In this case, these lessons were learned through YouTube videos and online tutorials. Nick often spoke of the the-atrical nature of Alinea, and all of the different elements (from software, to food, to serviceware, to kitchen layout) that were designed to create the desired outcome. It strikes me that James seems to share his father's passion for learning how things work, learning the intricacies of various software programs and technologies, and coordinating the people, locations, and props needed to pull off the production, which in this case, is a film rather than dining.

Makers Share Their Time and Tools

As a parent, few things are more painful than watching something happen that undoes your child's hard work and threatens to ruin a moment that you know they are looking forward to. At Bay Area Maker Faire 2012, this happened to a young attendee, and the outcome is a story (*http://bit.ly/maker-story*) that spread rapidly throughout the faire and the maker world.

Adam is a young maker who designed and built a go-kart during a summer program, and had the chance to show it off at Bay Area Maker Faire. Unfortunately, when the go-kart got to the faire, one of the bolts holding the steering wheel in place was missing a nut. If there's any place on the planet where you're likely to be able to find the part you need, Maker Faire is that place. Thus, Adam set off looking for the right sized nut. Amazingly, despite the large number of makers and equipment at the faire, no one had the correct piece. At this point, many people would give up, or duct tape the steering wheel in place. Not Adam, and not at Maker Faire. Instead, Adam went on a quest to get the part he needed. He spoke to makers at a booth for computer-aided design software and they sprung into action helping him design a new nut. Next step? Sending the file to a 3D printer (a common sight at Maker Faires!). A little while later, Adam had a custom-made, red plastic nut that fit his go-kart's bolt perfectly.

While this example took place at a Maker Faire, I am happy to say that it doesn't take a faire to pull together this sort of group "can do" spirit. Walk into a local makerspace and I can almost guarantee that you will see makers helping each other with their projects.

When the Maker Education Initiative, "Maker Ed," launched its national youth-serving Maker Corps program in 2013, we had a staff of four people. Part of the project involved packing more than 150 boxes with project materials. Maker Ed didn't have any storage space (other than my basement) in Minnesota, but we sud-denly had thousands of batteries and markers, hundreds of toothbrushes, MaKey MaKey, Squishy Circuit kits, construction paper, tape, glue, and other bits and pieces piling up. Our local makerspace, the Mill, donated storage space and helped organize groups of volunteers to help us pack the boxes that would be shipped to makers throughout the United States for Maker Corps' training program. Fueled with pizza and fruit, local makers from Minneapolis and St. Paul donated their time to pack and seal boxes. Even our local UPS store pitched in, picking up our boxes while waiving the pickup fee. The UPS employees even took a tour of the maker-space before heading off with the boxes.

Teachers Everywhere

Judy Aime' Castro (Figure 8-2) is a maker who educates both children and adults in a wide variety of making disciplines, such as electronics and sewing. Within moments of meeting Judy, her passion for sharing her skills and empowering others as makers is apparent.

Figure 8-2. Judy Aime' Castro leading a Teach Me To Make workshop (photo courtesy of Judy Aime' Castro)

Judy grew up in Peru, at a time when terrorism and turmoil made formal schooling difficult. As a child whose parents valued education, there was never any doubt that Judy would be taught. When there was no third grade, people in the neighborhood took it upon themselves to teach the children. As Judy put it, *"Education came from different people in our neighborhood. ...Education was not necessarily coming from a school setup, but as long as you create an environment for people to learn, that is an education. So it could be coming from different people, grandpa, grandmom, the neighbor."* Whoever was willing to teach became the teacher, and wherever they taught became the classroom. In this way Judy learned woodworking, kite-building, and sewing. The children in her neighborhood would come together and make their own toys and, when necessary, they'd find people in the community to teach them what they needed. They would teach each other the skills they needed.

It doesn't seem far-fetched to state that Judy's parents kindled their daughter's maker instincts. Her father was a machinist who liked to tinker and take electronics apart. It was he who taught Judy how to hotwire a car, which turned out to be useful when she was a 16-year-old living in New York. *"My first car when I was 16 and living in NY... I got this little beat up car, didn't have a starter, the key ignition was broken, my dad taught me how to connect everything... everything was a hack in that car... I learned to change the oil, everything I could."*

When the other students were going to school in an unofficial uniform of jeans and similar clothing, Judy went to school in *"pantaloons and skirts and blue hair."* As the daughter of a seamstress, Judy was skilled at making her own clothes, a skill that gained her attention. Other students were in disbelief that somebody could actually make a skirt or pants. At a time when *"it was all about how short the miniskirts were,"* this was a useful skill. Judy recalls the *"first time I taught some of the popular kids how to shorten their skirts in the bathroom. We didn't have a classroom and didn't want anyone to know."* Ironically, the high school Judy and her classmates attended had home economics classes. However, *"these were the popular girls. They would never go to home ec and get caught in the classroom."*

Interestingly, Judy also avoided the home economics class. She, however, avoided it because she already knew how to sew and had set her sights on the school's shop and auto repair classes. *"I wanted to take shop with the guys and learn how to do repairs and I remember my counselor said they don't allow girls in that class.... I was put in music class instead. I was very upset. I had guy friends who took the class and I'd go and visit them. Any pretense to get to work with the machines and tools."* These days, no pretense is needed for Judy to work with machines and tools. With Michael Shiloh, she founded Teach Me To Make, an educational outreach program. She has traveled the world teaching workshops for adults and children. The little girl who, when faced with the prospect of no formal schooling, used her community as a classroom, is now teaching the next generation of children how to see the world as full of possibility. Her work focuses particularly on outreach to underserved communities, and multilingual workshops. At the end of her discussion with me about her life path, Judy reflected:

> *I think it is essential for anybody to start working with their hands as a kid [so that] it doesn't become so foreign later on. As adults I think we lose that intuitive way of looking at things. We were playing more [as children]. I think that learning how to make things at a later age makes it a little more difficult. It's something you have to work twice as hard to do. Whereas if it's something that is encouraged, or in my case I was allowed to do what I wanted to do without any supervision, I don't think I would have had the same love for what I do if I had to start only after high school, trying to figure out what I wanted to do in life. I just simply followed what I intuitively felt was right.*

Teach What You Know

I've given many talks and led many workshops for audiences big and small. That said, I can honestly say that I was pretty nervous about the Squishy Circuits workshop at the 2013 World Maker Faire. Why? Because I wasn't the lead instructor. The lead instructor was my five-year-old daughter. When I offered her the chance to attend her second Maker Faire with me, she jumped at the chance and was surprisingly enthusiastic when I made the offhand suggestion that she give a talk. Growing up in a house where both parents have spent time as educators, and having been tagging along to Squishy Circuit talks and workshops since she was three months old, Sage (Figure 8-3) didn't need to think long before agreeing. She told me she wanted to apply to give a workshop so that she "could teach kids how to build circuits so that they can teach their parents." With that, and about 20 pounds of homemade play dough, we were off to New York.

So on a sunny day in September, my daughter and I found ourselves in a tent at the New York Hall of Science at a table with participants ranging in age from about 5 to about 70. Like many a first-time teacher, she found that sound was an issue. If you've ever been to a Maker Faire, you've noticed how loud it is. If you're five, and somewhat nervous, it's not that easy to be heard. On the spot she decided to skip her prepared comments and jump into the project. She handed out materials and, after asking me to loudly give a few instructions, walked around the table for the next 45 minutes helping the students one-on-one. Yes, she occasionally made mistakes. In one case, a woman of about 20 whose circuit wasn't working looked at me and said, *"but I did exactly what she told me to do."* In retrospect, just the fact that this adult was taking circuit-building instruction from a five-year-old was rather exciting. We quickly realized that my daughter had given her some insulating, rather than conductive, material. However, at one point I noticed her sitting next to a boy about her age, both of them smiling when the circuit he built lit up. His father voiced amazement that his son would listen to this young teacher when he often wouldn't take instructions from his father.

Figure 8-3. Sage Thomas leading a Squishy Circuits workshop at the 2013 World Maker Faire (photo courtesy of Margot Vigeant)

In that moment, I found myself thinking of Judy Aime' Castro's observation that *"Education was not necessarily coming from a school setup, but as long as you create an environment for people to learn, that is an education."* At the end of the workshop, another maker, this one a 12-year-old boy who was presenting work elsewhere in the faire, came to the tent to talk to my daughter. He wanted to check in with her on how her workshop went and to congratulate her. Seeing these two kids, neither one yet a teenager, talking about their excitement for teaching others was the highlight of my faire weekend.

Everyone, Everywhere, a Possible Teacher

The current Maker Movement exists because of the connections people are making and the ideas they are sharing. Whereas once people might have gathered at sewing bees or coffee shops to share knowledge, limited by who in their area had access to tools or information, we live in an age where we can almost instantly, digitally, connect with others around the world who share our interests or have the knowledge that we're looking for. As a parent, this is exciting because it means that we don't need to be the "expert." If your child wants to learn something new, there is likely someone in your community, or on an online forum, who has done something similar. Project sites, some specifically for children, allow exploring what others have done to be a simple undertaking. Detailed tutorials and instructions for countless projects are posted on curated sites. Even better, these sites provide an outlet for you and your children to post the things you've made. Many manual skills, though, are best learned in person rather than through the Internet. Similarly, typing back and forth with someone isn't quite the same as being in the same physical place at the same time, touching the same tools and materials. This, I believe, is why gatherings like Maker Faires are growing in both number and size. This is also why we're seeing fix-it clinics opening up in libraries and makerspaces in recreation centers, and libraries making tools available for their patrons to check out.

Generosity is something best learned through example. Nathan Seidle learned about electronics through the generosity of others on computer bulletin boards. Nick Kokonas watched his grandfather pursue his passion for violin making and then volunteer that skill to help others in his community. Judy Aime' Castro grew up seeing everyone around her as a possible teacher who she could learn from. All three of these makers grew up to follow those examples. If children are taught to fear that others will copy them, or that successful people are too busy to help beginners, would we expect them to grow into adults who are generous with their time, knowledge, and belongings? It seems like a better way is to show them the power of collaboration, and in helping others, and to let them observe us thanking those who are generous toward us.

9

Optimism

Makers are optimistic.

They believe that they can make a difference in the world.

It's easy to find pessimism. News sites often read like laundry lists of the most horrible things that have happened, are happening, or will happen. A combination of social, economic, medical, and environmental crises presented one right after another have a tendency to make people feel frightened, frustrated, and powerless. Children are introduced early to the scary parts of life. My kindergartner knows what to do if a "bad person with a gun" enters her school thanks to mandatory lockdown drills, and I have to admit that my policy of letting her listen to unlimited NPR broadcasts while on car trips looks, in retrospect, like it was a bad idea.

Apparently I'm not the only parent who struggles with helping their children develop their sense of the current state of the world. In his essay, "The Future Will Have to Wait," (*http://bit.ly/fwhtw*) Michael Chabon notes that his *"[eight-year-old] son seems to take the end of everything, of all human endeavor and creation, for granted. He sees himself as living on the last page, if not in the last paragraph, of a long, strange and bewildering book."* When I read Chabon's essay, I kept coming back to these two sentences. When I think about what I most want for my children, I realize that it's optimism. I know this is not the same as blind faith that the future will be great, but I believe they can feel an informed optimism where they understand the problems that the world faces and still feel hope that people, including themselves, are good and skilled enough to change things for the better. I want them to have the sense that their place in that *"long, strange and bewildering book,"* has pages and pages both before and after it. I want them to learn from the choices, expertise, and actions of those who came before them, and to have the confidence to play a role in writing the future for those who come after them.

Empowerment

A word that many of us use when we discuss the Maker Movement is "empowerment." At its core, empowerment involves feeling like you can make a change in the world around you, that you are capable. This resonates with makers, because we are reminded that the most impressive projects are completed only when we understand how to use the tools available to us. To become empowered, people must first believe in themselves and in their abilities, and that they are capable of creating things that have never been created before—be it a drastically new form of transportation, or a personalized dress that uniquely represents their sense of form and function.

Every maker that I spoke to mentioned a person—a relative, a mentor, a teacher, a sibling, a friend—who took the time to support them and to believe in them. These people cared enough to encourage them, sometimes in small ways. For these young makers, that encouragement often made the difference between powering past small mistakes to finish a product and giving up entirely. They persevered because their project mattered to someone. Choosing to mentor, or raise, a child is choosing to take a chance on the future. This is the seed of optimism, and the presence of hope and optimism is one of the things that makes the Maker Movement so powerful, particularly to children.

When I am around makers, I am struck by the absence of *can'ts*. Be it at a faire or a makerspace, I am surrounded by people discussing and working on projects that seem overwhelmingly ambitious, maybe even a little crazy. But makers don't overly concern themselves with why these projects shouldn't be taken on; instead, their energy is focused on how to make the improbable a reality. I've been amazed to watch people volunteer to pitch in, or advise, or fund, or just observe and cheer on, the undertakings. When we look at the early stages of most of the great innovations of humankind, from vaccines to telephones to space flight, we see lots of laughing and pooh-poohing. It's these wild, crazy, seemingly impossible ideas and projects, though, that have led to many of humankind's greatest breakthroughs. And I see the Maker Movement rallying around these ideas, nurturing the wild and hopeful and dreamy aspirations of their fellow makers.

From the first time that I attended a Maker Faire, I was intrigued by the people who were drawn to participate. As the previous chapters have discussed, there are some traits that seem to be common among makers. However, there's something more. Something that made me want to spend time around them. This was a can-do group of people who for the most part saw the people and the world around them in a positive light. When Dale Dougherty, Maker Faire's founder, talks about the faires he often stresses how optimistic he feels as he walks around looking at the incredible variety of projects. Even better, there's a tangible sense that attendees are actively developing the belief "that they can do things." Makers believe that they can make a difference in the world. This doesn't mean that all makers are setting out to cure diseases or to launch a space shuttle from their backyard. What it does mean is that this is a group of people who believe they can learn new skills and bring their ideas into a tangible form. Makers young and old are willing to put the time and effort in to create things that are personal to them, even when it might be easier (or cheaper) to buy something off the shelf.

Designing for the Far-Off Future

If raising children is a biological way of showing hope for the future, then perhaps the Clock of the Long Now is a technological way of expressing that hope. While some engineers spend their careers trying to create ways to get products to last for decades, the team behind this clock is implementing a design intended to endure for 10,000 years. This clock, currently under construction in a mountain in Texas, is envisioned as a way to inspire long-term thinking. To set out to build a device like this requires two fundamentally optimistic beliefs: first that we can create something that is more durable than anything anyone has ever created before and second, that there will be someone around to witness the clock 10,000 years into the future. The optimism reflected in these beliefs is not just about the durability of the product being made, but in the durability of the kinds of people who would tackle these endeavors.

The optimistic maker behind the Clock of the Long Now is Danny Hillis (Figure 9-1). When he introduced this project in 1995, or 01995, as members of the Long Now Foundation (*http://bit.ly/mill-clock*) would refer to it, he wrote, "*[I] cannot imagine the future, but I care about it. I know I am a part of a story that starts long before I can remember and continues long beyond when anyone will remember me. I sense that I am alive at a time of important change, and I feel a responsibility to make sure that the change comes out well. I plant my acorns knowing that I will never live to harvest the oaks.*"

Danny's own childhood was full of adventure. His father was a virologist and his mother a biostatistician who studied hepatitis. This meant that their children, Danny and two siblings, grew up following hepatitis outbreaks throughout Africa. A young Danny spent time in Rwanda, Burundi, the Belgian Congo, Kenya, and India. The family moved to the Congo in the middle of a civil war and all of their possessions were stolen en route, so the Hillis children found themselves in a new place with few distractions. While living abroad, they usually had little access to technology or books in English, so when he did get his hands on a book he devoured it and the knowledge in it. He also pursued the opportunity to learn new things with a rare tenacity. While living in Calcutta, Danny's mother convinced the British Consulate to allow him, about 11 years old at the time, to use their library (though they did make him read the books on site, rather than bring them home.) He threw himself into Boole's *Laws of Thought*, even though he was, in his own words, *"way too young to understand the book."* He was able to grasp the basic idea of Boolean Logic, and even at this young age realized that computers would allow him to make things that could think.

Figure 9-1. A young Danny Hillis (center) at a science fair he organized at his school in India (photo courtesy of Argye Hillis)

One of the joys of childhood, for Danny, was that it was a time when we believe that we can do anything, because we don't know what is impossible. He reflected that *"one of the toughest things as an adult is that you learn how hard things are so you don't try to do some of the things that were the best thing you thought of, because you know how impossible it would be."* In contrast, children truly believe that you can build a rocket ship, or a computer, out of found objects and enthusiasm. He became enamored of science fiction after being introduced to it by a favorite school librarian, who gave him a copy of *The Wonderful Flight to the Mushroom Planet*. Kids in this book built a rocket, which inspired Danny to start building, and ultimately exploding, rockets. As he read more science fiction, he became influenced by the characters in the stories. One of his favorite books at that time was *Have Spacesuit, Will Travel*. The young hero of this book saves the world, which leads to a girl falling in love with him and a scholarship to MIT. Young Danny Hillis had never heard of MIT at that point, but the idea of saving the world and falling in love sounded great, so MIT likely was wonderful, too. From that point on, he told everyone who asked that he would be going to MIT one day. And he did. He even ended up teaching there.

Over many conversations with Danny, I've wrestled with whether optimism is the best word for what he exudes. While he's the first to agree that he is an optimist, it seems that gratitude, rather than optimism, is his driving force. Every story that he tells includes something that someone taught him or helped him with. Danny is acutely aware of the opportunities that he has had throughout his life. Having parents who studied epidemics meant that he and his siblings met people for whom life was a struggle. As a small child he once heard his mother complain about something, and told her that the situation wasn't fair. Her reply? *"Life's not fair, and you're lucky it's not."* Even at a young age, Danny was aware of many possibilities that were open to him, but not to the other children he met, based on the circumstances of his birth. This meant that the Hillis children grew up in a family that was grateful, and that didn't take their opportunities for granted.

Not all of Danny's projects are on a multi-millenial timeline. Like many of the makers in this book, Danny is a parent (Figure 9-2). His children, who I discussed in Chapter 2, are all skilled makers. Perhaps this isn't surprising, because their father is a renowned inventor of many high-tech devices and a pioneer of parallel computing. All three children express the same gratitude and awe about the world that is so evident in their father.

Figure 9-2. Danny Hillis with his sons, Noah and Asa, in a tree house that the family built (photo courtesy of Asa Hillis)

Perhaps this is the truest way of expressing and perpetuating hope in the future: raising children who do the same. While the child at the beginning of this chapter may envision a world marching toward certain demise, Danny writes (*http://bit.ly/ d-hillis*) that he *"cannot believe that we are at the end of this story—we are not evolution's ultimate product. There's something coming after us, and I imagine it is something wonderful. But we may never be able to comprehend it, any more than a caterpillar can comprehend turning into a butterfly."*

Taking Action

Having hope for the future isn't always easy, particularly for those who know a lot about how things are made. To call Dawn Danby an optimist would be an oversimplification. This 36-year-old mother, and Senior Sustainable Design Program Manager at Autodesk, has a more nuanced view of the future, a view that is consistent with how she saw the world as a child. As a designer, educator, and parent, she has never been afraid to work hard for causes that she believed in.

As a child, Dawn spent as much time as possible outside and thinks fondly about *"being 12 and having big posters of the Carmanah Valley in British Columbia and becoming vegetarian, wearing anti-fur buttons and [giving] speeches about protecting dolphins."* In her high school environmental science class she was taught about climate change, nuclear reactors, OPEC (Organization of the Petroleum Exporting

Countries) and its politics. Dawn took this exposure for granted, not realizing how few teenagers were exposed to these topics at that time. She convinced her parents to take her on a trip over old logging roads so that she could see the place pictured in the poster on her childhood wall: one of the last untouched watersheds.

The beauty of natural places had a strong effect on her. Dawn studied art intensely, traveling internationally to do so. This led her to attend the Rhode Island School of Design (RISD), where she assumed she'd become an artist. It didn't work out that way. The turning point in her academic career came from a homework assignment. She was given a 20 × 30 piece of Bristol board and was challenged to turn it into an enclosed form with no paper left over. Even now, nearly two decades after working on that assignment, she still speaks of *"the feeling of sitting in my dorm room suffering and the feeling of the gears locking in. The project seemed impossible, since it meant having to think very precisely in three dimensions. You could almost feel the synapses firing."* Dawn acknowledges that if she had chosen a career by pursuing things that gave her pleasure, she would have become a painter or singer. Rather, she *"went into [industrial design] because the field of [industrial design] made me so angry. Because the amount of waste generated by consumer products made me so angry."* It was clear that industrial design was going to be a challenge for her, full of frustrations and contradictions, but it was a field where she felt she could make a difference. *"I just feel like the paradoxes and the contrasts keep increasing... The more amazing green tech, and solutions we see, the more the electronic companies are gluing in their batteries and making it harder to change the system."* Her exposure to beautiful forests and meadows as a child inspired her to preserve those areas as an adult by rethinking the systems that seemed to threaten her cherished wide open spaces.

After graduation from RISD, Dawn lived off of the grid working for the Center for Maximum Potential Building Systems and later as a medical illustrator, using visuals to explain complicated concepts. Today, Dawn works with designers and educators who are creating tools to teach, and implement, the principles and practices of sustainability for engineers and architects. When I speak to her about how we're educating the next generation of makers, she admits that she is worried that we aren't instilling students with a basic knowledge of how the planet works. She regularly meets young engineers who don't seem to realize how their designs will impact the world around them. As a woman who grew up constantly aware of how the built world impacts the natural world, and vice versa, she is passionate about sharing this knowledge.

In 2013, Dawn's daughter Meridian was born. I asked her whether she was optimistic about the future that Meridian will live in:

My daughter is going to grow up in wild times. All I know is that we can't see, and we can't know, what she will experience or what her generation will experience. I can provide her with an ability to manage uncertainty and see lots of things, and know lots of people. No content I provide will be up to date—what's important is that she knows how to creatively solve problems as they come up. Optimistic, pessimistic? Life will be complex. In some places there will be incredible innovation, in others there will be apocalypses. We know this because that is what's happening now.

With her knowledge of the natural and physical world, as well as the social factors that impact both, Dawn is raising Meridian in a household infused with the maker mentality. Living outside of San Francisco, in a part of Oakland that is weathering transitions, her family is growing food in the yard of the 1855 house that they are retrofitting. She and her neighbors tend to a secret urban chicken ranch where they are raising chickens, eggs, and vegetables. She reflects that her neighborhood's strength is that it truly is a community, where *"people say hi to each other on the street."* Dawn sees opportunities in the Maker Movement. *"As we see big systems break down, we see this interesting emergence of people making for themselves. We want to have a generation of people who are confident enough to devise the solutions when life is disrupted or the power runs down."*

Do It Yourself, Use What You Have

Sarah Grudem and Molly Black (Figure 9-3) are two makers devising solutions to environmental problems that they see around them. These two women can often be found at Sassy Knitwear, a small store in Minneapolis full of brightly colored clothes handmade largely from recycled apparel. In an era where it is easy to find cheap, mass-produced clothing that is current for the fashion whims of the moment, these two women take a different approach. Their company was catalyzed by Molly's concern about how much clothing was going into landfills from thrift stores. Rather than simply be worried about it, she launched a business that made clothes in a way that had a substantially lower environmental impact. She doesn't see this as a sacrifice, or even as a costly option. To Molly, her business model is the only feasible business model: *"It's the responsible thing to do. There's no option. You either do it this way or not at all."* Their conviction for creating clothing, indeed, for forming a business that is in keeping with their beliefs about how the world *could be* is emblematic of the attitude that I see among maker entrepreneurs.

They're not concerned with doing things the way things are traditionally done, but rather doing them the way that they feel things should be done.

Sarah and Molly both grew up in maker households. Their parents met through a parent-child music class when the girls were three, and the two families helped found a Waldorf preschool. From a young age, the girls were exposed to the idea that people made things, fixed things, and used things that other people had made. Sarah's grandmother earned a degree in home economics from Purdue in 1943. After living through the Depression, she instilled in her children and grandchildren a sense that if something is needed, it can be made. This particularly resonated with a young Sarah, who found freedom in these notions of self-sufficiency and avoidance of waste. Mostly self-taught in sewing, she took it upon herself to learn the skills she needed, and summed up her approach as *"Do it yourself. Use what you have."*

Figure 9-3. Sarah Grudem, Molly Black, and Christine Delmonico; Molly's mother crimped the girls' hair, and Sarah and Christine's mother made their dresses (photo courtesy of Sarah Grudem)

Sarah's dad was an elementary school art teacher who instilled a passion for art in his twin daughters teaching them about mixing paint colors even before they could talk. While the family had a TV for the first few years of the girls' life, a childhood tantrum when they were five years old led to their father putting the TV on the picnic table, inviting Molly over, and helping all of the kids take the TV apart, never to be repaired. (True to the family's belief that things should be used and reused, the TV's case ended up being the frame for the family's Christmas card picture, with the girls "on screen" as shown in Figure 9-4, with doves on their heads.) Tools were always around, and their dad built them a toy made out of circuit parts that could be reconfigured. Over at Molly's house, Molly's mother was the one who fixed things in their home. She also taught the girls how to sew and paint and garden.

Figure 9-4. Sarah Grudem, age nine, inside of the TV that she helped her father dismantle (photo courtesy of Anne Johnson)

Through the eighth grade, Molly and Sarah continued their schooling in Waldorf schools, which had a strong emphasis on handwork and seeing learning through a creative lens. They also reflected that the Waldorf model of students having the same teacher from grades 1 to 8 was powerful, and gave them a mentor.

Molly and Sarah are driven and confident, and their business, right down to the name Sassy Knitwear, reflects their bold attitudes and willingness to invest themselves in their work. Between them they have no formal background in clothing design or business, but they are doing everything they can to learn and grow while maintaining a successful venture. When they needed more sewers, they hired local women, some of whom they had to train how to sew. When they realized that they'd need more fabric colors, they sourced eco-friendly fabrics and learned how to hand dye with low-impact dyes.

The interior and bright colors of Sassy Knitwear (Figure 9-5) invites adults to spend time browsing, and mixing and matching styles, but it's also inviting to kids, with a smattering of toys in one corner and Sarah performing essential functions with a sleeping baby in her arms. Unsurprisingly, their children are learning to make things for themselves. Sarah's four-year-old helps her mother pick fabrics and Molly's eight-year-old son is so enthused about making clothing that he also makes things for his sister and others, and recently designed and sewed his own Halloween costume. These children are growing up in a family with a long history of making and community involvement. It doesn't seem a stretch to imagine them sharing their mothers' optimism and belief that doing things the right way is the only way to do them.

Figure 9-5. Sarah and Molly celebrating the Sassy Knitwear store's first birthday (photo courtesy of Sarah Grudem)

I have been amazed by the generosity of the makers who were willing to give me hours of their time and share intimate details about their lives. With no exceptions, I left every coffee meeting, phone call, Skype interview, and random encounter feeling re-energized. These are people who love what they do and the people they do it with.

Jim Henson, maker extraordinaire, and father to the Muppets, *Sesame Street*, and numerous advances in the field of puppetry, once said *"When I was young, my ambition was to be one of the people who made a difference in this world. My hope is to leave the world a little better for having been there."* This is the sentiment that I hear over and over again in my conversations with makers. Each, in his or her own way, is making the world a little better by empowering someone else, or designing a new tool, or by making someone smile in delight at the things they've made.

Optimism is a word that is often thrown around almost casually. People are quick to label others as optimists or pessimists. I believe that it's more nuanced than that. Optimists can find themselves frustrated, they can lash out about the injustice they see, they can even take an idea that they have poured their passion into and throw it out the window. But optimists don't leave it at that. They tend to be driven to continue trying, to take a new approach, or to recruit help from others because they know that their efforts are important, that they have real consequences. The optimism that we see exhibited by the makers, young and old, in this book and elsewhere, is about a belief that all of us have the capacity to, as Henson put it, *"make the world a little better for having been there."*

10

Maker Moms/Dads/ Teachers/Neighbors/ Friends

Susan was a mother whose kids came to her asking for a sled. They wanted one that was faster than the neighbors' sleds. Instead of saying no, she spoke with her children about how sleds are made. Then they built one. Susan had been best in her class at math and sciences and, as a child, had spent lots of time helping out in her father's carriage shop. She was exactly the sort of parent who could help her kids build a fast sled. Susan was the family member who fixed things and made things, and coached her children on building projects. It's not surprising that her sons became makers. They had a little company selling toys to the neighbor kids: they made kites and they sold those. As teenagers, they built a letter press, they built a lathe, and I'd like to think that some of the lessons they learned from their mother about designing a sled came in handy when they made their own transportation device later in their lives.

You may have guessed, correctly, that the brothers in the previous paragraph are Orville and Wilbur Wright (*http://bit.ly/wright-bros*), inventors of the first airplane. My laptop is covered with stickers, and my favorite one says *"Maker Mom: Mother of Invention."* If ever there was a woman who exemplifies that title, it's Susan Wright (*http://bit.ly/susan_wright*). Describing his childhood, Orville Wright said that *"We were lucky enough to grow up in an environment where there was always much encouragement to children to pursue intellectual interests; to investigate whatever aroused curiosity. In a different kind of environment, our curiosity might have been nipped long before it could have borne fruit."* (*http://bit.ly/wright-pdf*) Over and over, I have heard similar sentiments from the makers I spoke to. They have immense gratitude for the people around them (*http://bit.ly/kat_wright*) who let them explore their interests as kids. I heard stories of neighbors, parents, teachers, and friends who encouraged the passions and pursuits of the children in their lives. In every maker's life story, there was at least one moment when the support of someone else convinced them to keep working and to keep making. What follows are the lessons that I have taken away from the amazing stories I've heard.

Share Your Passions

Whatever it is that you love to make, be sure that the children in your lives get to see that passion. Even if they don't find themselves interested in that pursuit, they will see that it's OK for adults to be excited about making things and learning new skills. I was amazed by how detailed makers' recollections were of what members of their families, and their communities, made—from neighbors who would occasionally do craft projects with them, to parents who spent their free time tinkering in the basement. In some cases, I suspect these influencers weren't even aware that their actions were being noticed, such as the case of the neighbor one maker vividly recalled watching from afar as he worked in his garage. Even more powerful were the situations where the young makers were invited into the workshops and garages to watch and participate.

When I visited Sassy Knitwear to interview Sarah Grudem and Molly Black (Chapter 9), I was delighted to see how seamlessly their children were integrated into the fabric of their shop and their sewing. Sarah's infant son was nestled in a sling on her chest as she spoke with me and did tasks around the store. Kids' toys could be found in a basket close at hand, and all of the children are growing up watching their mothers design and sew clothing. The older children join in by making their own clothes and costumes. Making, and making together, is a common occurrence. Who knows if Sarah's and Molly's children will still be drawn to sewing when they're adults. However, they will be able to look back on days spent creating their own clothing with their mothers and watching an idea turn into a thriving business.

Dawn Danby, the sustainable design expert and industrial designer, brings her young daughter with her as she tends to gardens, and chickens, in her neighborhood. Meridian's smiling face can be seen peeking over her mom's shoulder as she rides in a backpack. Whether she remembers these early nature walks or not, she will surely see pictures of herself visiting neighbors, and chickens, and vegetable gardens (Figure 10-1), and know that this was something her mother believed to be important enough to involve her in from a very early age.

Figure 10-1. Dawn tending community gardens with her daughter Meridian (photo courtesy of Dawn Danby)

Let Children Follow Their Own Interests

Very few, if any, of the makers in this book ended up in careers that closely followed those of their parents. Most spoke fondly of relatives and mentors who encouraged them, but all forged their own paths. In some families, it took years for the makers to feel as though their choices were respected, particularly when those choices strayed dramatically from the schools or careers their parents had hoped they would pursue. Now parents themselves, some makers find it frustrating when their own children fail to be interested in the same tools and technologies as they are. One father described his unsuccessful attempts at trying to get his children to share his excitement about 3D printing and other projects that he took on in his basement.

Supporting your children in an endeavor that you also enjoy is easy. It's when your children find themselves fascinated by projects that you personally have no interest in that your encouragement is truly tested. I have been, and continue to be, thrilled when my daughters' share my interests in sewing, electronics, coding, baking, and woodworking. However, when my oldest daughter became intrigued by collecting animal skeletons, dead bugs, and feathers, it took every ounce of my patience not to ask her to find something else to do. I've never been a fan of dead bugs, and was less than ecstatic when Sage left a pungent fish vertebrae in my office. I had to admit, somewhat begrudgingly, that the bone's geometry was fascinating. I also had to apologize to my research students and colleagues about the dead fish smell radiating from my office for days. I took a deep breath—or in the case of the fish, a very shallow breath—and found other people who could enthusiastically support her in this new interest. My father-in-law has been generous in providing her with animal skeletons and wasps' nests that he found on his rural property, and my husband took on the task of learning how to preserve the severed bird wing that Sage found on her school playground.

We don't get to pick our children's interests, but we do get to influence how broad an array of experiences they are exposed to. Most importantly, we get to choose how we encourage the endeavors and interests that they choose for themselves.

Step Back

We've all heard stories about parents helping a bit too much with homework or projects. I'll never forget the time I walked into a university machine shop to find a professor intensely focused on the part he was turning with the lathe. Upon closer inspection, it became obvious that the part had little to do with the precise equipment that he used in his lab. When I asked what he was working on, he somewhat sheepishly admitted that it was his son's Pinewood Derby car. He really wanted the child to do well, but felt that a bit of extra help would be needed, and so he had brought the car into lab for a tune-up, without his son.

I kept this experience in mind when my daughter Sage, then four years old, was working on a Nerdy Derby car to bring to World Maker Faire in New York. The Nerdy Derby, a variation of the traditional Pinewood Derby race, allows entrants to build their cars out of anything, with virtually no requirements other than that the car fit on the track. I bought a simple starter kit and let my daughter put it together and decorate the car however she wanted. As more and more decorations were added, I had to resist the urge to tell her that it was likely her decorations would slow the car down. It was her car, not mine, and if she wanted buttons on the wheels and glitter... everywhere, so be it (Figure 10-2). So what if the aesthetics aren't quite what I would have done, or the aerodynamics are poor? A slow car that she made all by herself would ultimately mean more to her than the fastest car that I could have made for her. These days, when she and I sew together, I keep quiet (unless asked) when it comes to fabric choices and pattern lengths. Decisions like these are hers, not mine, and I think of the makers in this book, such as Jane Werner and Sophi Kravitz, who boldly designed their own styles from an early age and how this has given them confidence throughout their lives. And then I smile when Sage picks the loudest patterns and brightest ribbons with which to attire herself.

Figure 10-2. Sage Thomas with her Nerdy Derby car, at the 2012 World Maker Faire, wearing a dress she helped sew

Every maker needs to find his or her own voice. One of the greatest gifts that we can give them is to step back and let them discover it for themselves.

Teach the Importance of Safety and Responsibility

As I wrote this book, I would send drafts to the person whose story I was telling to make sure that I was getting it right and that I wasn't sharing anything that they wanted to keep anonymous. The most sensitive stories, as I discussed earlier, were almost always the ones in which a potentially dangerous incident was being discussed. Sometimes I left these out of the book, at the request of the maker, or kept the story anonymous. Why? There were two reasons that were most often given for the *"please don't share this with my name"* requests. One was *"my parents don't know about this."* The other was *"I don't want my kids to know that I did this."* More than once, I had interviewees tell me that after we spoke they called up their parents to find out why they let them do the crazy things that they did as a kid.

Holly Gates, who I wrote about in Chapter 7, commented that:

> I'm sure today [my parents] would be considered highly negligent. But I think it benefited me, helping me to learn to do things on my own, to identify and pursue my own interests and agenda.

Holly's father, Bob, sees it differently:

> [I] thought allowing, encouraging Holly to explore and experiment on his own was better for Holly's development than over-parenting, which would deprive Holly of many useful experiences and life lessons. For instance, he touched the cabin heater and burned his hand, but did not touch it again, and gained more awareness of heat energy and proximity....parents were not negligent.

This is a nice illustration of a topic that all of us who support young makers need to think about: when do we jump in and stop kids from trying something? At what point are the risks too great?

One maker had a very clear recollection of a parent stepping in. John Edgar Park, Director of Digital Production & Technology at DisneyToon Studios, was interested in electricity and explosives from a young age. His parents supported his endeavors, to a point. In seventh grade, John convinced his mother to write a check for some supplies that he wanted. Little did she know, he was ordering large quantities of fireworks. When the package arrived and his father saw what it was, he made John watch as he poured water over them in the bathroom sink, reducing John's haul of powerful fireworks to a powerless wet mush. I asked John what his reaction was to watching his father destroy his purchase:

Gosh, I think my main takeaway at the time was that my Dad was very concerned about my safety, and with the vast amounts of great judgment I resolved to still acquire dangerous materials, but be sneakier about it and "safe" when using them so that I didn't prove him right and blow my face off. To be fair, my father told lots of stories of near injury doing fun, dangerous stuff when he was a kid, so I figured everything would turn out fine.

John's father was a teacher and high school administrator, a member of the U.S. Army and Navy Intelligence at various times, and later a diplomat, which meant that John traveled as a child, and lived for a year in Pakistan. As a teenager in Pakistan, John found that it was easy to get access to the chemicals and materials that he wanted for making mischief. Unfortunately, he also learned that 240V wall power is stronger than the 120V in the United States. This led to an incident in 10th grade where he was thrown a few feet across the room after shorting out an exposed transformer. Given that experiences like this could have led to dire consequences for him as a child, I was curious what he considered acceptable risk for his own children:

Acceptable risk is a tricky one, because to put your kids in harm's way kind of paints you as a monster. I want to protect them from serious injury or dismemberment, but I'm compelled to empower them to try interesting and useful things. I encourage them to do "dangerous" stuff such as using hot soldering irons (they helped me solder dozens of LEDs for a project when they were five and seven), sharp knives, and other pointy, pinchy hand tools. I'm so proud now watching my kids approaching new tools and skills with respect, but not fear. I reserve the right to reverse my decision on all of this the first time I find out they're secretly building highly unstable explosives like I did as a kid.

In situations where serious danger isn't imminent, I find myself remembering Christy Canida and her daughter Corvidae's ability to cook and orienteer at the age of four. As part of a family that is avidly involved in outdoors and construction endeavors, Corvidae has been taught from a young age how to assess risk and determine her capabilities. Perhaps most importantly, she has been gaining new skills alongside her family and other skilled mentors. Just like her mother, she'll likely grow up with quite a few bruises and scars, but also an appreciation for tools and their safe usage. In my own family, I have allowed my daughters to use real tools from a young age. Every tool, be it a sewing machine or a hammer, has also come with training and supervision in its use.

Let Kids Get Their Hands, Clothes (and Even Their Rooms) Dirty

Play and learning can be messy, often literally. Fortunately, children and houses are washable. Many of the projects and adventures described in this book involved messiness. One parent described his maker child as being *"mostly naked and covered in mud and paint"* during his preschool years. Many adult makers, particularly those who attend Burning Man, still spend periods of time thus adorned.

Will Durfee, the engineering professor whose mother let him work on telescopes in his room even though it entailed a 50-gallon drum, had this suggestion: *"Parents: let your kids build in their rooms! So what if a blob of solder puts a small burn into the carpet?"* While many parents might not be quick to accept a hole in the carpet, it is worth spending some time thinking about what boundaries you are comfortable with. In our house, we have an unfinished basement where (almost) anything goes. This is where we head to paint, glue, glitter, solder, and generally make a mess. My husband and I, as well as our daughters, do our best to clean up afterward, but if some permanent paint gets on the floor we're not going to think twice about it. Part of being a maker is learning to respect your tools and your environment, lessons that you are never too young to learn. Once kids know where it is OK to be messy, and where it is not, making becomes a lot less stressful.

A large portion of my daughter's body was tinted green for about a week after a preschool painting project went awry. At the time, I was upset, but she eventually returned to her normal color, still loves to paint, and we have a family story (and pictures) that make us all laugh. The paint was nontoxic, and she had a grand time doing it. I don't necessarily recommend that other kids change their hue, but the incident was a good reminder, for me, that most things are fixable, and the chance to explore and be a bit silly is often worth the clean up.

You Don't Have to Have All of the Answers

You don't need to be an expert to support young makers. No parent, teacher, or mentor could ever be expected to be knowledgeable about every aspect of most projects. Rather, a strength of makers is their ability to find and connect with people who have the knowledge and skills that are needed for the project at hand. I've met many parents and teachers who helped the kids in their lives find other community members who could mentor them.

Recently, as I was flying to an event where I would be speaking about young makers, I absentmindedly checked my Facebook feed and saw the following post from my friend Eric:

> *Dear Dad: Thank you for buying me a copy of Stacker for Christmas, way back when I was 11. I know you didn't really understand what it did, and I know $100 was a lot for the family to spend on a Christmas present that year. And I know how mad you must have been (even though you didn't show it) when I decided to delete the large 100 MB "compressed volume" file on the ghost D drive to make even more space, destroying the past six months of your work.*

I, like Eric's father, had no idea what Stacker was, so he kindly explained it to me. *"Back in the day, hard drives were of course super-small (ours was 106 MB) but you could run these 'disk compression' programs that would hook into DOS (ugh) and compress and decompress the data on the fly. This of course made things a little bit slower but I guess everything was slow back then, and hey, an extra 100 MB! (Average compression ratio was about 2x.) They did this by creating a fake D: drive that held the compressed volume via the real OS. DOS was of course incredibly primitive so this was hacks on top of hacks, and sometimes a dumb kid like me could get confused by the compressed volume and whoops!"*

The post was written by Eric Jonas, an engineer and scientist who co-founded Prior Knowledge, a company that developed predictive databases and which raised more than a million dollars in seed funding before being acquired by Salesforce. I know Eric as the brilliant first-year student who lived down the hall from me in our dorm at MIT. Despite my being two years ahead of him at MIT, it was Eric I would turn to for help figuring out how to program in C or wire a circuit. Quickly upon moving into his freshman dorm room, Eric set up an impressive electronics lab that took over half the room, and could always be found soldering circuits late into the night. While Eric's father was able to see Eric graduate from MIT, he passed away before seeing Eric develop Prior Knowledge and earn his doctorate. The post I saw was written years after his father had died, and is just one of the many examples of how his parents supported Eric's interests even when they had very little knowledge of what exactly it was that he was doing. Whether this meant saving money to buy the computer program that Eric wanted, or resisting the urge to be outwardly angry when Eric's tinkering led to the loss of his father's work data, they encouraged him to try new things and ask for their help when needed.

Eric's story is a reminder that we don't always need to know the answers. It's OK to tell your kids, or your students, that you don't know how to do something.

In that moment, though, the most important thing is the *next* step. These moments of uncertainty are a chance to say "let's figure it out together," and head to the library, a makerspace, a project sharing website, or a neighbor's house.

Even better, go learn something with the young makers in your life. John Edgar Park, the maker whose father (and a kitchen sink) thwarted his youthful fireworks' explorations, can often be found doing projects with his two children. Recently, this included taking a machine sewing class with his daughter (see Figures 10-3 and 10-4. It's been fun to watch as this led to the duo using their new skills at home. By gaining a new skill together, or taking on a project that requires you to learn new information, you are demonstrating that you are never too old to learn something new, and that it's OK to admit that you aren't sure how to do something.

Figure 10-3. John Edgar Park and his daughter, Bea, with bags they made together at a sewing class (photo courtesy of John Edgar Park)

Figure 10-4. John and Bea working on a project at home (photo courtesy of John Edgar Park)

Now Go Make Something

Now for the easy part. To encourage the young makers in your lives, simply start making. Build a bench, wire up an LED lantern, grab a paintbrush, bake a pie... make something. Even better, try making something you've never made before. Being a maker is about the communal act of making and sharing, not the object you make. So find, or create, that community. If you believe, as I do, that we are all born makers, it becomes easy to find other makers. Opportunities, and teachers, are all around us.

The Makers in
This Book

Chris Anderson is the CEO of 3D Robotics and the founder of DIY Drones. Prior to that, he was the Editor in Chief of *Wired* magazine. He has a degree in physics and did research at Los Alamos. After initially failing out of college, he spent time in Washington, D.C., as a messenger and playing in a punk rock band.

Molly Black currently lives in Minneapolis with her husband and two children. She co-owns Sassy Knitwear, an ethically made, organic/sustainable clothing store. She can be found most days at the Sassy store creating clothing, often with one of her children "helping" out. When she is not at the Sassy Knitwear store or doing Sassy Knitwear–related work, Molly enjoys playing with her kids experimenting with new recipes, gardening, reading, and dancing.

Kipp Bradford is an entrepreneur, technology consultant, and educator with a passion for making things. He is the founder or cofounder of start-ups in the fields of transportation, consumer products, HVAC, and medical devices, and holds numerous patents for his inventions. Some of his more interesting projects have turned into kippkitts. He also spent some time playing glockenspiel in a community street marching band, and raced road bicycles semi-professionally for many years.

Leah Buechley is a designer, engineer, artist, and educator whose work explores intersections and juxtapositions—of "high" and "low" technologies, new and ancient materials, and masculine and feminine making traditions. She also develops tools that help people build their own technologies, among them the LilyPad Arduino kit. While an associate professor at MIT she founded and directed the High-Low Tech research group. Her work has been exhibited internationally in venues including the Victoria and Albert Museum, the Ars Electronica Festival, and the Exploratorium.

Christy Canida is the Senior Manager of Parterships at Instructables. She earned an S.B. in biology from MIT, and has worked in the biotech industry, academic research labs, and an aquarium. She is an avid cook and has posted Instructables for recipes randing from Sous Vide Chicken Thighs to Bacon Pixie Stix.

Judy Aime' Castro been making things all her life without being aware of how many other people shared her interest. Judy combined her knowledge, skills, and interests in co-founding Teach Me To Make, an art and technology educational and outreach program. Judy is also sole proprietor of Aime' Designs, a custom maker studio specializing in soft products. She has been involved in several collaborative art projects in the Bay Area, is a Tinkerer in Residence at the Exploratorium, and manages local events for Arduino.

Dawn Danby works across disciplines in sustainable design. She is passionate about finding ways to apply human ingenuity to eliminating ecological impacts, and is currently exploring ways to encourage makers to channel their creativity toward tackling planetary challenges. She created the Autodesk Sustainability Workshop, which teaches young engineers, designers, and architects the principles and practice of sustainable design. Dawn is currently working on coaxing her 130-year-old house toward being net-zero in energy consumption.

Lindsay Diamond was born and raised in Sarasota, Florida. She earned a doctorate in biomedical sciences from the University of Florida. Fueled by a passion for kinesthetic learning and the concept of play in her role as the Director of Education at SparkFun, Lindsay leads a team of educators and engineers developing resources for tinkerers and makers of all ages. When she was 27, Lindsay taught herself to sew by making Christmas stockings using hot pink fabric with skulls and crossbones, which ended up selling quite well!

Will Durfee is Professor and Director of Design Education in the Department of Mechanical Engineering at the University of Minnesota, Minneapolis. He received his undergraduate degree from Harvard and graduate degrees from MIT. His professional interests include medical devices, rehabilitation technology, product design, and design education. He has not yet figured out how to keep sawdust out of electronics projects in his basement maker shop.

Lenore Edman runs Evil Mad Scientist Laboratories, a small hobby electronics business, along with Windell H. Oskay. The two of them also blog about projects involving art, electronics, food, design, and whatever else piques their interest. Lenore once shared an office with Edgar Allan Poe's writing desk, but these days she shares her office with a variety of robots.

Woodie Flowers grew up in Jena, Louisiana. He earned his B.S. from Louisiana Tech and his M.S., M.E., and Ph.D. degrees from MIT, and went on to become a professor in MIT's Mechanical Engineering Department. Passionate about education, Woodie is the co-founder of the FIRST Robotics Competition, a member of the National Academy of Engineers, and the former host of *Scientific American Frontiers*. He loves tools. He once started listing the motors he owns and got bored and quit at 170.

Holly Gates is a photovoltaic engineer at 1366 Technologies, Inc., of which he is also a founding member. Previously, Holly has worked at E Ink and the MIT Media Lab. When not in his lab, he can be found doing things such as creating patterns for his family's clothes, canning and preparing food, or restoring old machinery. Holly's blog, Tooling Up (*http://tooling-up.blogspot.com*), chronicles these adventures.

Bradley Gawthrop is a pipe organ builder whose varied interests include photography, cycling, writing, electronics, security, printing, typography, cooking, and history. He lives with his wife (who he believes is significantly out of his league) and many cats, who are utterly unmoved by his technical and dialectical efforts but respond favorably to table scraps and scratches around the ears.

Sarah Grudem currently lives in Minneapolis with her husband and two young children. She co-owns Sassy Knitwear, an ethically made, organic/sustainable clothing store there. She also regularly performs classical harp music, with groups or as a soloist, from weddings to the Minnesota Orchestra. Her spare time is spent quilting, knitting, baking, reading, or gardening. She, and her twin sister Christine, have been friends with Sassy Knitwear co-owner Molly Black since they were three years old.

Asa Hillis grew up on a small farm in Southern California with his twin brother and younger sister. From early childhood and into his teens he was inspired to build and disassemble things. Currently he is putting this passion into earning a degree in furniture design at the California College of the Arts. Upon completion, he plans on working with his twin brother to design and specialize in custom furniture that reaches beyond the bounds of mass production.

Danny Hillis is an inventor, scientist, author, and engineer. He is co-founder of Applied Minds, a research and development company that invents, designs, creates, and prototypes high-technology products and services for a broad range of applications. Previously, Danny was Vice President and Disney Fellow, Research and

Development at Walt Disney Imagineering, and co-founder of Thinking Machines Corp. Danny is also co-chairman of The Long Now Foundation, Judge Widney professor of engineering and medicine at the University of Southern California, and adjunct professor of engineering at the USC Viterbi School of Engineering. He is the designer of a 10,000-year mechanical clock.

India Hillis grew up on a small farm in Southern California with her two brothers. She is currently studying product design at Art Center College of Design. Previously, she studied at the Central Saint Martins College of Design in London.

Noah Hillis grew up on a small farm in Southern California with his twin brother and younger sister. He is a student at the California College of the Arts, where he is pursuing a degree in furniture design. In his spare time, he restores vintage travel trailers.

Steve Hoefer was born and raised on a farm in Northern Iowa where he learned by doing, and has been learning ever since. He has been a forensic animator, beekeeper, video game designer, baker, and English teacher. Steve currently works as a writer and technologist, and spends his spare time fascinated by old patents, dreaming of being an astronaut, and taking things apart even though they work just fine.

Mimi Hui is a designer and engineer who spent her childhood in New York City and Macau. As the founder of Canal Mercer Designs, Mimi consults with companies ranging from start-ups to established companies on consumer interface and design challenges. Mimi earned a B.S. in electrical computer systems engineering from Rensselaer Polytechnic Institute, and an M.A. in innovation, design thinking, and industrial design from Brunel University. As a member of NYC Resistor, Mimi has worked on many whimsical projects, including an electronic piano made of JELL-O. She aspires to be a polyglot and is known to rise with the birds to practice languages.

Jeffrey Jalkio is an associate professor of engineering and physics at the University of St. Thomas. As a graduate student, he cofounded CyberOptics, a manufacturer of noncontact metrology systems. These days, his research focuses on measurement uncertainty and how to deal with it. In his spare time, Jeff can be found practicing T'ai Chi, reading, or attending theater in the Twin Cities.

Steve Jevning, the founder of the Leonardo's Basement youth makerspace, built his first big project when he was five years old: an airplane on a huge fallen elm tree. The tree was trimmed and the top branch formed a two-person cockpit onto which a four-foot-long 2 × 4 prop was nailed. Two summers ago, he helped a group of teens build a half-sized replica of a Beechcraft King Air. Some dreams fly forever.

Eric Jonas is a scientist, engineer, and entrepreneur who likes building things at the intersection of electrical engineering, machine learning, and neurobiology. He earned a Ph.D. in neuroscience from MIT, as well as an S.B. in brain and cognitive sciences, and B.S. and M.E. degrees in electrical engineering and computer science. He was the CEO and co-founder of Prior Knowledge, Inc., and the Chief Predictive Scientist at Salesforce.com. Eric grew up in Boise, Idaho.

Dean Kamen is the founder and president of DEKA Research & Development Corporation. Examples of technologies developed by DEKA include the HomeChoice™ portable dialysis machine, the iBOT™ Mobility System, the Segway™ Human Transporter, a DARPA-funded robotic arm, a new and improved Stirling engine, and the Slingshot water purifier. Kamen has received many awards for his efforts, including the National Medal of Technology in 2000 and the Lemelson-MIT Prize in 2002. He was inducted into the National Inventors Hall of Fame in 2005 and has been a member of the National Academy of Engineering since 1997. Among Kamen's proudest accomplishments is founding FIRST (For Inspiration and Recognition of Science and Technology), an organization dedicated to motivating the next generation to understand, use, and enjoy science and technology.

David Kelley is the founder and chairman of the global design firm IDEO and the founder of Stanford University's Hasso Plattner Institute of Design, also known as the d.school. Stanford's Donald W. Whittier Professor in mechanical engineering, Kelley influences the field of design by pioneering human-centered design methodology, the culture of innovation, and design-thinking education. Kelley is a member of the National Academy of Engineering. His numerous awards include the Robert Fletcher Award from Dartmouth College, the Edison Achievement Award, Chrysler Design Award, National Design Award in Product Design from the Smithsonian's Cooper-Hewitt National Design Museum, and the Sir Misha Black Medal for his "distinguished contribution to design education."

Nick Kokonas is the co-founder and owner of Alinea Restaurant, Next, and the Aviary. He pioneered the use of tickets and variable pricing for the restaurant industry. Nick has a degree in Philosophy from Colgate University. He then worked as a derivatives trader, and founded his own trading firm, prior to partnering with chef Grant Achatz in opening Alinea Restaurant.

Sophi Kravitz has been making stuff since she was a little kid. She is a formally trained engineer whose first career was in physical special effects props for film. It was then, while creating works that were seen on small screens or large, that she realized the great fun in creating works that can easily satisfy an audience or group of participants. She currently has a technical consultancy where she helps clients with research, electronics, and systems design. In addition to designing electronics and machines, Sophi has made six wedding cakes.

Allison Leonard designs firmware and hardware as an electrical engineer at MakerBot Industries. She grew up in Oregon where she spent a lot of her time exploring the outdoors. Allison enjoys exploring machines by taking them apart, which she documents at her website, the Machines Project (*http://machinesproject.com/*).

Luc Mayrand's experience includes projects in themed entertainment, film, television, and games, in the United States, Japan, England, France, and Korea. He initially worked as creative director and concept designer with independent producers, with projects for Universal, Tri-Star, Paramount Parks, Herschend, Sony, Sanrio, Samsung, etc. Since 1998, Luc has been with Walt Disney Imagineering, as executive creative director and show producer on several E-ticket projects, as well as ongoing Blue Sky work. For the past six years, Luc has been a core lead in the development of Shanghai Disneyland for China. He is a fan of the ultimate adventure hero: Bob Morane, created by French writer Henri Verne, and owns copies of the first 150 novels.

Paul McGill is an electrical engineer at the Monterey Bay Aquarium Research Institute, where he designs robotics and instrumentation for ocean research. This work has taken him to the South Pole, within 1,500 kilometers of the North Pole, and down 4,300 meters to the bottom of the ocean. After high school, Paul served in the U.S. Air Force, where he gained practical electronics skills while maintaining F-15 fighter jets. He then went on to earn electrical engineering degrees at Stanford University.

Amon Millner, a visiting assistant professor of computing innovation at Franklin W. Olin College of Engineering, develops tools that make it possible for people to capture happenings in the physical world and map them to applications running on computers. He was on the design team that produced the Scratch programming language. Amon has a Ph.D. in media arts and sciences from MIT, an M.S. in human computer interaction from Georgia Tech, and a B.S. in computer science from USC. To his family's great amusement, he once repaired a fixer-upper car that only drove in reverse when he bought it.

John Edgar Park is a maker, animation filmmaker, technical/artistic problem solver, TV host, writer, builder of quirky electro-mechanical contraptions, optimist, husband, and father of two. He is the Director of Digital Production & Technology at Disneytoon Studios, overseeing CG filmmaking and technology research and development on such films as Disney's *Planes*, *Planes: Fire & Rescue*, and the *Disney Fairies*. John hosted the American Public Television show *Make: Television* and has presented talks at many Maker Faires. He writes for *Make:*, Boing Boing, Cool Tools, and other places online and in print, including authoring the book *Understanding 3D Animation Using Maya* (Springer). John's obsessions include coffee roasting/brewing/making/drinking, robots, tools, beatboxing, sharing self-portraiture, and admittedly strange human body tricks.

Lisa Regalla is Interim Executive Director for the Maker Education Initiative. Before joining the Maker Ed team, she was the Manager of Science Content & Outreach at Twin Cities Public Television where she was responsible for the educational content presented on television, in person, in print, and on the web as part of the Emmy Award-winning series, *SciGirls* and *DragonflyTV: Nano*. Lisa previously served as an educator at the Museum of Science, Boston, and the Da Vinci Science Center in Pennsylvania. Lisa received both a B.S. in chemistry and a B.A. in theater from Lehigh University before earning her Ph.D. in chemistry from the University of Florida. Outside of work, Lisa's passion is participating in endurance events with Team in Training to raise money for the Leukemia & Lymphoma Society.

Mitch Resnick, professor of learning research at the MIT Media Lab, develops new technologies and activities to engage people (especially children) in creative learning experiences. His Lifelong Kindergarten research group developed ideas and technologies underlying the LEGO Mindstorms robotics kits and Scratch programming software, used by millions of young people around the world. He also co-founded the Computer Clubhouse project, an international network of 100 after-school learning centers where youth from low-income communities learn to express themselves creatively with new technologies. Resnick earned a B.S. in physics from Princeton, and an M.S. and Ph.D. in computer science from MIT. He was awarded the McGraw Prize in Education in 2011.

Luz Rivas started her career at Motorola, where she was an electrical design engineer working on position and navigation systems for the automotive industry. As an educator, Luz has developed programs focused on increasing underrepresented minorities in STEM fields at Caltech and most recently was a director at Iridescent, where she worked on training programs for engineers interested in teaching kids. In 2011, Luz founded DIY Girls. She has a B.S. in electrical engineering from MIT and a masters in technology in education from the Harvard Graduate School of Education.

Eric Rosenbaum is a doctoral student in the Lifelong Kindergarten group at the MIT Media Lab, where he continues his quest to expand your imagination. Eric develops new technologies for creative learning, but also encourages others to make creative mouth noises, invent their own board games, and play. He has degrees in psychology and educational technology, and plays the funky trombone.

Nathan Seidle is the founder and CEO of SparkFun Electronics, which he started out of his dorm room in college. Nathan is active in the open hardware community, and serves on the board of the Open Source Hardware Association.

Raquel Vélez is a Senior Software Developer at npm, Inc. in Oakland, California. She studied mechanical engineering at the California Institute of Technology (Caltech) and worked as a roboticist for eight years at a variety of institutions, including the NASA Jet Propulsion Laboratory, the University of Duisburg-Essen, the MIT Lincoln Laboratory, the University of Genoa, and Applied Minds. Raquel speaks five languages, is fairly certain laughter and chocolate will cure everything, and uses robots as an excuse to get people excited about code and math.

Jane Werner is the executive director of the Children's Museum of Pittsburgh. She studied synaesthetic art education at Syracuse University, and has worked on things as varied as constructing exhibits, editing film, raising money, sewing wedding dresses, and restoring an 1819 farmhouse.

Index

We'd like to hear your suggestions for improving our indexes. Send email to index@oreilly.com.

Acknowledgments

This book would not have been possible without the generosity of the many makers who let me interview them for this book. In addition to the makers that you've met on the previous pages, I spoke to many others who shared stories from their childhoods. Thank you all!

The team at Maker Media has helped me at every step of this process, sharing stories, references, and advice. Brian Jepson has been an amazing guide through the process of writing a book. Thank you to Dale Dougherty and everyone in the Maker Media and Maker Education Initiative teams—you have all strongly influenced my understanding of the Maker Movement.

I am very grateful to those who read early versions of this book and gave thoughtful feedback, notably Colin Angevine, Katherine Acton, Deb Besser, and John Chernega. Thank you to my colleagues and students at the University of St. Thomas for always furthering, and challenging, my thoughts about education and learning. In particular, I am thankful to Don Weinkauf for providing a space for faculty to pursue their ideas, even when it means building a lab in a hallway or putting engineering students on a flying trapeze.

My husband Chris has been supportive of this project before it was even a book. Thank you, Chris, for always being there, no matter how ambitious, ridiculous, or unlikely the project! Thank you to my family, for their support and patience. My parents, Jack and Reggie Polsenberg, have always encouraged me in my endeavors, no matter how messy or unusual. My children, Sage and Grace, put up with my long hours on the computer while writing this book, and are truly the inspiration for this book.

About the Author

 **Dr. AnnMarie Thomas** is an associate professor in the School of Engineering at the University of St. Thomas. Her teaching and research focus on engineering design, particularly as it relates to PK–12 education. With her students, AnnMarie explores the playful side of engineering on topics such as Squishy Circuits, the Engineering of the Circus, and toy design. Prior to joining the faculty at the University of St. Thomas, she taught at Art Center College of Design. AnnMarie served as the Founding Executive Director of the Maker Education Initiative, and she is the mother of two young makers. AnnMarie has an S.B. in ocean engineering (with a minor in music) from MIT, M.S. and Ph.D. degrees in mechanical engineering from Caltech, and a professional certificate in sustainable design from the Minneapolis College of Art and Design.

Colophon

The cover image is by Kristian Olson. The cover fonts are URW Typewriter and Guardian Sans. The text font is Adobe Minion Pro; the heading font is Adobe Myriad Condensed; and the code font is Dalton Maag's Ubuntu Mono.